S0-BWX-434

THE ARITHMETIC THEORY OF QUADRATIC FORMS

By
BURTON W. JONES

THE
CARUS MATHEMATICAL MONOGRAPHS

Published by
THE MATHEMATICAL ASSOCIATION OF AMERICA

THE CARUS MATHEMATICAL MONOGRAPHS are an expression of the desire of Mrs. Mary Hegeler Carus, and of her son, Dr. Edward H. Carus, to contribute to the dissemination of mathematical knowledge by making accessible at nominal cost a series of expository presentations of the best thoughts and keenest researches in pure and applied mathematics. The publication of the first four of these monographs was made possible by a notable gift to the Mathematical Association of America by Mrs. Carus as sole trustee of the Edward C. Hegeler Trust Fund. The sales from these have resulted in the Carus Monograph Fund, and the Mathematical Association has used this as a revolving book fund to publish the succeeding monographs.

The expositions of mathematical subjects which the monographs contain are set forth in a manner comprehensible not only to teachers and students specializing in mathematics, but also to scientific workers in other fields, and especially to the wide circle of thoughtful people who, having a moderate acquaintance with elementary mathematics, wish to extend their knowledge without prolonged and critical study of the mathematical journals and treatises. The scope of this series includes also historical and biographical monographs.

The following monographs have been published

No. 1. Calculus of Variations, by G. A. Bliss

No. 2. Analytic Functions of a Complex Variable by D. R. Curtiss

No. 3. Mathematical Statistics, by H. L. Rietz

No. 4. Projective Geometry, by J. W. Young

No. 5. A History of Mathematics in America before 1900, by D. E. Smith and Jekuthiel Ginsburg

No. 6. Fourier Series and Orthogonal Polynomials, by Dunham Jackson

No. 7. Vectors and Matrices, by Cyrus Colton MacDuffee

No. 8. Rings and Ideals, by Neal H. McCoy

No. 9. The Theory of Algebraic Numbers, by Harry Pollard

No. 10. The Arithmetic Theory of Quadratic Forms, by B. W. Jones.

The Carus Mathematical Monographs

NUMBER TEN

THE ARITHMETIC THEORY OF QUADRATIC FORMS

By

BURTON W. JONES

University of Colorado

Published by

THE MATHEMATICAL ASSOCIATION OF AMERICA

Distributed by

JOHN WILEY AND SONS, INC.

Composed and Printed
By
The Waverly Press
Baltimore, Maryland
1950

INTRODUCTION

The arithmetic theory of quadratic forms may be said to have begun with Fermat in 1654 who showed, among other things, that every prime of the form $8n + 1$ is representable in the form $x^2 + 2y^2$ for x and y integers. Gauss was the first systematically to deal with quadratic forms and from that time, names associated with quadratic forms were most of the names in mathematics, with Dirichlet playing a leading role. H. J. S. Smith, in the latter part of the nineteenth century and Minkowski, in the first part of this, made notable and systematic contributions to the theory. In modern times the theory has been made much more elegant and complete by the works of Hasse, who used p-adic numbers to derive and express results of great generality, and Siegel whose analytic methods superseded much of the laborious classical theory. Contributors have been L. E. Dickson, E. T. Bell, Gordon Pall, A. E. Ross, the author and others. Exhaustive references up to 1921 are given in the third volume of L. E. Dickson's History of the Theory of Numbers.

The purpose of this monograph is to present the central ideas of the theory in self-contained form, assuming only knowledge of the fundamentals of matric theory and the theory of numbers. Pertinent concepts of p-adic numbers and quadratic ideals are introduced. It would have been possible to avoid these concepts but, in the opinion of the author, the theory gains elegance as well as breadth by the introduction of such relationships. Some results and many of the methods are here presented for the first time. The development begins with the classical theory in the field of reals from the point of view of representation theory, for in these terms many of the later objectives and methods may be revealed. The successive chapters gradually narrow the fields and rings until one has the tools at hand to deal with the classical problems in the ring of rational integers.

The analytic theory of quadratic forms is not here dealt with because of the delicate analysis involved—in fact, the theory of the singular series of Hardy and Littlewood plays such a role in this theory that much of analytic number theory would be a prerequisite for such discussion. However, some of the more important results are stated and references given.

In most cases no attempt is made to credit ideas to their discoverers since, for the most part each idea is partly the property of several persons. However, in some clear cut cases, sources are acknowledged specifically. The author owes many debts to Siegel and Pall, whose works he has studied in detail. Much of the work on this monograph was done while the author was on Sabbatic Leave from Cornell University and with the help of a grant from the Research Corporation; thus the author is indebted to both organizations.

The theory of quadratic forms is rather remarkable in that, though much has been done, in some directions the frontiers of knowledge are very near. It is hoped that this book will not merely be read with interest but stimulate explorations of the unknown.

Burton W. Jones

The University of Colorado

TABLE OF CONTENTS

Chapter I

FORMS WITH REAL COEFFICIENTS

1. **Fundamental notions.** Though in this chapter we shall be dealing primarily with quadratic forms with real coefficients, most of the definitions carry over into later theory and, by way of introduction, the questions we shall ask and answer will, for the most part, have their analogues later in the book.

Of fundamental importance is the notion of congruence which appears not only in this chapter but in our treatment of forms with rational and p-adic coefficients and which is closely allied to the notion of equivalence. Let

$$f = \sum_{i,j=1}^{n} a_{ij} x_i x_j \quad \text{and} \quad g = \sum_{i,j=1}^{n} b_{ij} y_i y_j$$

be two quadratic forms with real coefficients and $a_{ij} = a_{ji}$, $b_{ij} = b_{ji}$ for all i and j. If there are real numbers t_{ij} such that, when the values of x given by

$$(1) \qquad x_i = \sum_{j=1}^{n} t_{ij} y_j, \qquad i = 1, \cdots, n,$$

are substituted in f, the form g results, we say that the linear transformation (1) with real coefficients *takes f into g*. This transformation is called *non-singular* if (1) may be solved for y_j. We call f and g *congruent* forms and write $f \sim g$ if there is a non-singular linear transformation with real coefficients taking f into g. The importance of this notion stems from the fact that two congruent forms represent the same numbers (that is, take on the same values as the variables take on all real values) since for any number N, a set of values of y_i making $g = N$ will by (1) yield a set of values of x making $f = N$ and the transformation (obtained by solving (1) for y_j) taking g into f will take a solution of $f = N$ into one of $g = N$. Furthermore, there is a $1 - 1$ correspondence between the sets of values of x_i making $f = N$ and the sets of values of y_j making $g = N$. Thus, since we are concerned with the numbers represented by the forms, out of a set of congruent forms only one need be considered just as in geometry all equilateral triangles of unit side are the same as far as the properties we are interested in are concerned.

We define the *matrix of a form* to be the matrix A whose elements in order are a_{ij}, that is $A = (a_{ij})$, and the determinant of the form to be the determinant of A, which we denote by $|A|$. If we denote the row matrix $(x_1, x_2, x_3, \cdots, x_n)$ by X^T (we use the superscript T to denote the "transpose" of a matrix) and the corresponding column matrix by X, matric multiplication shows that we may write

$$f = X^T AX.$$

Then, since (1) may be written in matric notation in the form $X = TY$ where T is the matrix whose elements are t_{ij}, we recall that $X^T = Y^T T^T$ and see that the transformation (1) takes f into

$$g = Y^T T^T ATY = Y^T (T^T AT) Y$$

which shows that the matrix B of g is equal to $T^T AT$.

For example, if f is the form $2x_1^2 + x_1 x_2 + 3x_2^2$, its matrix is

$$A = \begin{bmatrix} 2 & \frac{1}{2} \\ \frac{1}{2} & 3 \end{bmatrix}$$

and

$$f = (x_1 x_2) \begin{bmatrix} 2 & \frac{1}{2} \\ \frac{1}{2} & 3 \end{bmatrix} \begin{bmatrix} x_1 \\ x_2 \end{bmatrix}.$$

The transformation $x_1 = 3y_1 + y_2$, $x_2 = 2y_1 - y_2$ takes f into $g = 36y_1^2 - y_1y_2 + 4y_2^2$ since

$$\begin{bmatrix} 3 & 2 \\ 1 & -1 \end{bmatrix}\begin{bmatrix} 2 & \frac{1}{2} \\ \frac{1}{2} & 3 \end{bmatrix}\begin{bmatrix} 3 & 1 \\ 2 & -1 \end{bmatrix} = \begin{bmatrix} 36 & -\frac{1}{2} \\ -\frac{1}{2} & 4 \end{bmatrix}.$$

Notice that if (1) can be solved for y_j we have in matric notation $Y = T^I X$ and $A = T^{TI} B T^I$ where we denote the inverse of a matrix by the superscript I.

If A is an n by n matrix and B an m by m matrix we call C the *direct sum* of A and B if

$$C = \begin{bmatrix} A & 0 \\ 0 & B \end{bmatrix},$$

where 0 stands for a zero matrix of appropriate size, and use the notation $C = A \dotplus B$. Furthermore, if f, g and h are the forms whose matrices are C, A and B respectively, we write $f = g + h$. In the direct sum of two or more forms we omit the dot above the plus sign since we agree to add two forms thus only if the variables in the one are distinct from those in the other. In an analogous manner we speak of the direct sum of more than two matrices or forms.

2. **A canonical form for congruence.** Since we need to investigate only one of a set of congruent forms, it behooves us to select as simple a form as possible from such a set. This is done by means of the following theorem which, with the exceptions noted, holds for forms with coefficients in any field (for a definition of a field see section 7). Since we shall need them later for more general fields, we shall prove them in general form. The reader not at home with more abstract fields may prefer first to consider the proofs in this section and the next only for real numbers.

THEOREM 1. *Every form* $f = \sum_{i,j=1}^{n} a_{ij}x_ix_j$ *with coefficients in a field in which* $2 \neq 0$, *may be taken into a form*

$$g = b_1y_1^2 + b_2y_2^2 + \cdots + b_ry_r^2$$

by a linear transformation whose coefficients are rational functions of the coefficients of f and whose determinant is $+1$.

To prove this theorem we use two different kinds of transformation. The first is of the type

$$x_1 = y_i, \qquad x_i = y_1, \qquad x_j = y_j, \qquad j \neq 1, i.$$

This interchanges the i-th and first rows and columns of A, the matrix of f, and leaves unaltered any minor determinant containing elements of the first and i-th rows and columns of A. The second is of the type

$$x_i = y_i + ry_j, \qquad x_k = y_k, \qquad k \neq i.$$

This adds r times the i-th row of A to the j-th and r times the i-th column to the j-th, and leaves unaltered the value of any minor determinant containing elements from the i-th and j-th rows and columns of A. The transformations of each type have determinant ± 1.

Now if $a_{ii} \neq 0$ for some i, the first transformation takes f into a form whose leading (that is, first) coefficient is not zero. If $a_{ii} = 0$ for all i let $a_{ij} \neq 0$ for some fixed $i \neq j$. First interchange the first and i-th rows and columns, then the second and j-th to have a form $f' = \sum_{i,j=1}^{n} a'_{ij}x_ix_j$ in which $a'_{12} \neq 0$. Then use the second type of transformation above with $i = 2$, $j = 1$, $r = 1$ which yields a form

$$h = \sum_{i,j=1}^{n} c_{ij}\, y_i\, y_j = (c_{11}\, y_1 + \cdots + c_{1n}\, y_n)^2/c_{11} + g$$

with $c_{11} \neq 0$ and g a form in $y_2, y_3, \cdots, y_n$. Then use the transformation

$$z_1 = (c_{11}y_1 + c_{12}y_2 + \cdots + c_{1n}y_n)/c_{11}$$
$$z_i = y_i, \qquad i > 1,$$

to take h into $c_{11}z_1^2 + g'$ where g' is obtained from g by replacing y by z. This last transformation may be recognized as the application of several transformations of type two; i.e. adding c_{1j}/c_{11} times the first row to the j-th and similarly for columns.

Now g' is a form in $n - 1$ variables and we can deal with it as with f. Since each of the transformations has determinant ± 1 with elements rational functions of the coefficients of f, the same is true of their product which is a transformation taking f into the desired form. If the transformation has determinant -1 we may change the sign of one of the variables without altering g and have a transformation of determinant $+1$.

The number r in the form g of the theorem is equal to the rank of the matrix of f since the transformation taking f into g is non-singular and hence does not alter the rank of A.

If now we permute the coefficients of g if necessary so that the first i coefficients are positive and then use the transformation

$$\sqrt{|b_j|}\, y_j = z_j, \qquad j = 1, 2, \cdots, r$$

we have proved

THEOREM 2. *Every quadratic form f with real coefficients is congruent to a form*

$$h = z_1^2 + z_2^2 + \cdots + z_i^2 - z_{i+1}^2 - \cdots - z_r^2 .$$

COROLLARY 2. *Every form is congruent to a form whose determinant is not zero.*

We call h the *canonical form* of f under a real transformation. In theorem 2, i is called the *index* of f and of its matrix. The *signature* of f is $2i - r$. These definitions are justified by

THEOREM 3. *If the forms*

$$f = x_1^2 + x_2^2 + \cdots + x_i^2 - x_{i+1}^2 - \cdots - x_r^2$$

and

$$g = y_1^2 + y_2^2 + \cdots + y_j^2 - y_{j+1}^2 - \cdots - y_s^2$$

are congruent, then $i = j$ *and* $r = s$.

First $r = s$ since the ranks of two congruent forms must be equal. Then we suppose that $j > i$ and arrive at a contradiction. Let T be the transformation taking g into f, that is

$$y_k = t_{k1}x_1 + t_{k2}x_2 + \cdots + t_{kr}x_r . \tag{2}$$

In this set of r equations put $x_1 = x_2 = \cdots = x_i = 0$. Then the equations $y_{j+1} = 0, \cdots, y_r = 0$ become $r - j$ homogeneous equations in the $r - i$ variables $x_{i+1}, \cdots, x_r$. But, since $r - j < r - i$, there are real values of $x_{i+1}, \cdots, x_r$ not all zero satisfying these equations. Since these values of the y's and x's satisfy (2), $g = f$, that is

$$-x_{i+1}^2 - \cdots - x_r^2 = y_1^2 + y_2^2 + \cdots + y_j^2$$

with $x_{i+1}, \cdots, x_r$ not all zero. This is impossible for real values of the x's and y's.

Similarly we can show a contradiction if $j < i$ and establish our theorem.

Furthermore, if two forms f and F have the same index and rank they will be congruent to the same canonical form h of theorem 2. If transformations T and S take

f and F respectively into h, then the transformation TS^I takes f into F and hence these two forms are congruent. Thus we have

COROLLARY 3. *Two forms are congruent if and only if their ranks and indices are equal.*

If a form f in n variables has index n, it is called a *positive definite form* or merely a *positive form.* If $i = 0$ and $n = r$ it is called a *negative definite form* or merely a *negative form.* These two types of forms are included in the term *definite form.* All other forms are called *indefinite.*

3. **The determination of the index of a form.** While one can get the index of a form by putting it into canonical form, there are more elegant ways of finding the index.

Let D_i be the i by i determinant in the upper left hand corner of a symmetric matrix A, that is, the leading i-rowed determinant of A. A matrix and its form are said to be *regular* if no two successive D_i are zero. We shall show that every non-singular symmetric matrix is congruent to a regular matrix and that the index of the matrix A depends on the signs of the D_i in the regular matrix. We first need two lemmas on matrices.

LEMMA 1. *Let A be an n by n symmetric matrix and A_{ij} the matrix obtained by crossing out the i-th row and j-th column of A. Then, defining D_i as above,*

$$|A|D_{n-2} = |A_{n,n}||A_{n-1,n-1}| - |A_{n,n-1}|^2$$

provided $D_{n-2} \neq 0$. This lemma holds for A a matrix with elements in any field.

If K is the matrix consisting of the first $n - 2$ rows of A, the fact that $D_{n-2} \neq 0$ shows that the last two columns of K are linear combinations of the first $n - 2$ columns. Hence appropriate linear combinations of the first $n - 2$

columns of A may be added to the last two columns and the same for rows to give a matrix

$$A' = \begin{bmatrix} A_0 & N & N \\ N^T & r & s \\ N^T & s & u \end{bmatrix}$$

where N is a zero column matrix with $n - 2$ elements, A_0 is the leading $n - 2$ by $n - 2$ minor of A, $|A_0| = D_{n-2}$ and r, s, u are numbers. Furthermore the process by which we got A' from A leaves unaltered D_{n-2} and the determinants of A_{ij} for $i, j > n - 2$. Our lemma then results from the equations: $|A| = |A'| = (ru - s^2)D_{n-2}$, $|A_{n-1,n-1}| = uD_{n-2}$, $|A_{n,n}| = rD_{n-2}$, $|A_{n,n-1}| = -sD_{n-2}$.

The theory of determinants tells us that if D is a non-singular r-rowed minor of a matrix and if every $r + 1$-rowed minor of the matrix is singular, then the rank of the matrix is r. Lemma 1 yields the following result for symmetric matrices

Lemma 2. *If D is a non-singular r-rowed principal minor of A and if every $r + 1$-rowed and $r + 2$-rowed principal minor of A containing D is singular, A is of rank r (A being symmetric). This lemma and the first part of its corollary hold for any field.*

From the result above quoted we need merely show that every $r + 1$-rowed minor containing D is singular under the conditions of this lemma. Choose any pair $i, j > r$ and let B be the matrix obtained from A by crossing out all its rows and columns except those of D and the i-th and j-th. Then, using B in place of A in lemma 1 we have

$$|B||D| = |B_{i,i}||B_{j,j}| - |B_{i,j}|^2$$

and $|B| = |B_{i,i}| = |B_{j,j}| = 0$ implies $|B_{i,j}| = 0$, which completes our proof.

This lemma shows that if A is of rank n, $k \leqq n - 2$ and D_k is non-singular, then by permuting the $k + 1, \cdots, n$ rows and columns of A (the same permutation for rows as for columns) we get a matrix whose D_{k+1} and D_{k+2} are not both zero. This yields the

COROLLARY. *Every symmetric matrix is congruent to a regular matrix. If the regular matrix is real and* $D_{k+1} = 0$, *then*

$$D_k D_{k+2} < 0.$$

Now we can prove the following result about the index of a form.

THEOREM 4. *The index of a regular symmetric matrix with real elements and of rank* n *is the number of permanences of sign in the sequence*

$$1, D_1, D_2, \cdots, D_n$$

where the D_i *are the leading i-rowed minor determinants of* A *and any zero* D_i *may be given arbitrary sign. (See the example below.)*

The theorem certainly holds for $n = 1$. We assume it for $n = r - 1$ and prove it for $n = r$. Let f_r and f_{r-1} be the forms whose matrices are the matrices of D_r and D_{r-1} respectively, let i_r and i_{r-1} be their respective indices and see that

$$f_r = f_{r-1} + 2 \sum_{i<r} a_{ir} x_i x_r + a_{rr} x_r^2.$$

As we have seen in the proof of theorem 1, f_{r-1} can be put into the canonical form of that theorem without altering the determinant of f_r. That is

$$f_r \sim \sum_{j=1}^{s} g_j x_j^2 + 2 \sum_{\substack{j \leqq s \\ k > s}} a_{jk}\, x_j x_k + \sum_{i,j>s}^{r} a_{ij} x_i x_j$$

$$= \sum_{j=1}^{s} g_j(x_j + a_{j,s+1}\, x_{s+1}/g_j + \cdots + a_{jr}\, x_r/g_j)^2 + \sum_{i,j>s}^{r} a'_{ij}\, x_i\, x_j,$$

where s is the rank of f_{r-1}. First if $D_r D_{r-1} \neq 0$ we see that $s = r - 1$ and $D_{r-1}a'_{rr} = D_r$. Thus if $a'_{rr} > 0$ the index of f_r is 1 more than the index of f_{r-1} and the pair $D_{r-1}D_r$ gives a permanence of sign while if $a'_{rr} < 0$ the indices are the same and $D_{r-1}D_r$ is not a permanence. If $D_{r-1} = 0$ then $D_{r-2}D_r < 0$ from the corollary of lemma 2 and $s = r - 2$. Let

$$h = \sum_{i,j=r-1,r} a'_{ij}\, x_i\, x_j$$

and see that $|h|\, D_{r-2} = D_r$ implies that $|h| < 0$ and hence that if h is put into the canonical form of theorem 1 it has one positive and one negative term. Thus the index of f_r is 1 more than the index of f_{r-2} and the sequence $D_{r-2}D_{r-1}D_r$ has one permanence of sign whatever the choice of sign of D_{r-1} may be. If $D_r = 0 \neq D_{r-1}$ then $D_{r+1} \neq 0$ and we can carry through the above process for $r + 1$ in place of r. This completes the proof.

Let us use this method to find the index of the matrix

$$\begin{pmatrix} 1 & 2 & 3 & 1 \\ 2 & 4 & 6 & 3 \\ 3 & 6 & 9 & 0 \\ 1 & 3 & 0 & 0 \end{pmatrix}.$$

Here $D_2 = D_3 = 0$ showing that the matrix is not regular. If then we interchange the third and fourth columns and rows we have the sequence 1, $D_1 = 1$, $D_2 = 0$, $D_3 = -1$,

$D_4 = 0$. Since there is one permanence of sign from 1 to D_1 and one from D_1 to D_3, the index of the matrix is 2 and its rank 3.

4. **Representation of one form by another.** We are now in a position to state and solve the general problem of representation for forms with real coefficients. (It is this problem of representation with which we are primarily concerned in this book.) Let f and g be two forms with n and m variables respectively and respective matrices A and B, $n \geqq m$. We seek conditions under which there will exist an n by m matrix T such that $T^T A T = B$. In such a case we say that the form f *represents* the form g. Notice that in determining the conditions for congruence we have already solved the problem for $n = m$ and that the case $m = 1$ will tell us under what conditions $f = N$ is solvable for N, a real number.

A preliminary theorem is the following due to Gordon Pall which, with an eye to future use, we prove for forms in any field.

THEOREM 5. *Let f and g be two forms with coefficients in a field, with n and m variables, non-singular matrices A and B respectively and $n > m$. Then there is a matrix T such that $T^T A T = B$ if and only if there is a form h in $x_{m+1}, \cdots, x_n$ such that f is congruent to $g + h$.*

First suppose f is congruent to $g + h$. Then there is a transformation T such that $T^T A T = B \dotplus C$ where C is the matrix of h. Write T in the form $(T_1 T_2)$ where T_1 is the matrix composed of the first m columns of T and T_2 of the last $n - m$ columns. Then

$$\begin{bmatrix} T_1^T \\ T_2^T \end{bmatrix} A(T_1\, T_2) = \begin{bmatrix} T_1^T A T_1 & T_1^T A T_2 \\ T_2^T A T_1 & T_2^T A T_2 \end{bmatrix} = B \dotplus C$$

and hence $T_1^T A T_1 = B$.

On the other hand, suppose there is a matrix T_1 such that $T_1^T A T_1 = B$. Recall that the rank of a product of matrices is not greater than the least of their ranks. Thus the rank of B, namely m, is not greater than that of T_1. But T_1 can have no greater rank than m since it has m columns and hence its rank must be m. Then by the lemma below we can find a matrix T_2 (with elements in the field) having n rows and $n - m$ columns such that $T = (T_1\ T_2)$ is non-singular. Let S be thematrix

$$\begin{bmatrix} I_1 & -B^I T_1^T A T_2 \\ 0 & I_2 \end{bmatrix}$$

where I_1 is the m-rowed identity matrix, I_2 is the $n - m$ rowed identity matrix. Then multiplication of matrices shows that S takes $T^T A T$ into $B \dotplus C_1$ where C_1 is non-singular since T and A are. Then if C_1 is the matrix of the form h in $x_{m+1}, \cdots, x_n$ we have f congruent to $g + h$. Notice that $TS = (T_1 - T_1 B^I T_1^T A T_2 + T_2)$ takes A into $B + C_1$ and that the first m columns of TS are T_1.

We used in the proof

LEMMA 3. *If T is an n by m matrix of rank m in a field, and $m < n$, then there is an n by $n - m$ matrix T_0 such that $(T\ T_0)$, the matrix whose first m columns are T and whose last $n - m$ columns are T_0, is non-singular.*

To prove this permute the rows of T to get a matrix T' whose first m rows are linearly independent, that is, which form a non-singular matrix. Then if T_0' is the n by $n - m$ matrix whose first m rows are all 0 and whose last $n - m$ rows form an identity matrix, the expansion of the determinant of $S = (T'\ T_0')$ shows that it is non-singular. Then if the permutation of rows taking T' into T is applied to S we have a matrix with the desired properties.

To return to the problem of this section suppose now that f and g are two forms with real coefficients and respective indices i and j. Then there is a matrix taking the matrix of f into that of g if and only if we can find a form h such that $f \sim g + h$. This immediately leads to the following theorem.

THEOREM 6. *A form f of index i and rank n represents a form g of index j and rank $m \leqq n$ if and only if $i \geqq j$ and $n - m \geqq i - j$; that is, if and only if $i \geqq j$ and $n - i \geqq m - j$.*

For the case $m = 1$ and $g = Nx^2$ we have

COROLLARY 6. *A form f with real coefficients, index i and rank n represents a real number N except in the following two cases:*

$$1.\ i = n, \qquad N < 0.$$

$$2.\ i = 0, \qquad N > 0.$$

A form represents zero non-trivially if and only if it is indefinite.

This last result can easily be derived independently of the theorem. For generalizations of these results see section 17.

5. **Automorphs and the number of representations.** For real matrices the existence of a solution X of $X^TAX = B$ implies the existence of an infinite number of solutions, but there is a relationship among the different solutions which it is worth while to develop. A transformation U is said to be an *automorph* of a form or its matrix A if $U^TAU = A$. It is obvious that $X^TAX = B$ and $X = UY$ with U an automorph of A implies that $Y^TAY = B$.

That for any field, all the solutions of $X^TAX = B$ can be derived from one in this fashion is shown by

THEOREM 7. *If A and B are two non-singular matrices in n and m variables respectively with $n \geqq m$, if their elements lie in a field with $2 \neq 0$ and if $X^TAX = Y^TAY = B$ then there is an automorph U of A such that $X = UY$.*

First if $n = m$ then $X^TAX = Y^TAY = B$ implies that X and Y are non-singular and thus $XY^I = U$ is an automorph of A and the theorem holds.

Second if $n > m$ we know that $X^TAX = B$ implies that there exists a non-singular $n - m$ by $n - m$ matrix C and a non-singular n by n matrix R such that $R^TAR = B \dotplus C$. Furthermore $X^TAX = B$ implies $(R^IX)^TR^TAR(R^IX) = B$. Now by the remark at the close of the proof of theorem 5 there is a transformation $(R^IX \; X_0)$ whose first m columns are R^IX taking R^TAR into $B \dotplus C'$ and similarly a transformation $(R^IY \; Y_0)$ taking R^TAR into $B \dotplus C''$. Now the next corollary below and $B \dotplus C' \sim B \dotplus C''$ shows that there is an $n - m$ by $n - m$ transformation S taking C' into C''. Thus the transformation $I \dotplus S$ (where I has $n - m$ rows and columns) takes $B \dotplus C'$ into $B \dotplus C''$ and $(R^IX \; X_0)(I \dotplus S) = (R^IX \; X_0S)$ takes R^TAR into $B \dotplus C''$. Thus, by the first part of this proof there is an automorph U' of R^TAR such that $(R^IX \; X_0S) = U'(R^IY \; Y_0)$. Thus $R^IX = U'R^IY$, $X = RU'R^IY$ and $RU'R^I = U$ is an automorph of A.

The theorem on which the above proof depends is

THEOREM 8. *Let $ax_0^2 + f$ and $ax_0^2 + g$ be two congruent forms in a field in which $2 \neq 0$ where f and g are quadratic forms in $x_1, \cdots, x_n$. Then $f \sim g$.*

To prove this theorem let

$$T = \begin{bmatrix} t_0 & t_1 \\ t_2 & T_0 \end{bmatrix}$$

be a transformation taking $ax_0^2 + f$ into $ax_0^2 + g$, where T_0 is an n by n matrix, t_0 is a number, t_1 is 1 by n and t_2 is n by 1. If A and B are the matrices of f and g, the matric equation

$$\begin{bmatrix} t_0 & t_2^T \\ t_1^T & T_0^T \end{bmatrix} \begin{bmatrix} a & 0 \\ 0 & A \end{bmatrix} \begin{bmatrix} t_0 & t_1 \\ t_2 & T_0 \end{bmatrix} = \begin{bmatrix} a & 0 \\ 0 & B \end{bmatrix}$$

is equivalent to the three equations

$$\begin{aligned} t_0^2 a &+ t_2^T A t_2 &&= a \\ t_0 a t_1 &+ t_2^T A T_0 &&= 0 \\ t_1^T a t_1 &+ T_0^T A T_0 &&= B. \end{aligned}$$

Choose the sign of $\pm 1 + t_0$ so that it is not zero, write $u = (t_0 \pm 1)^{-1}$ and define $S = T_0 - t_2 t_1 u$. Then

$$\begin{aligned} S^T A S &= (T_0^T - u t_1^T t_2^T) A (T_0 - u t_2 t_1) \\ &= T_0^T A T_0 - u t_1^T t_2^T A T_0 - u T_0^T A t_2 t_1 + u^2 t_1^T t_2^T A t_2 t_1 . \end{aligned}$$

Using the three equations we have

$$\begin{aligned} S^T A S &= T_0^T A T_0 + u t_1^T t_0 a t_1 + u t_1^T a t_0 t_1 + u^2 t_1^T (a - t_0^2 a) t_1 \\ &= T_0^T A T_0 + u t_1^T t_1 a \{ t_0 + t_0 + u(1 - t_0^2) \} \\ &= T_0^T A T_0 + t_1^T t_1 a = B. \end{aligned}$$

Hence $A \sim B$.

This theorem with theorem 1 establishes the

COROLLARY 8. *If A, B and C are symmetric matrices in a field in which $2 \neq 0$, then $C \dotplus A \sim C \dotplus B$ implies $A \sim B$.*

6. **Summary.** It is worth while to glance back over the methods and results of this chapter, since, to a certain extent, they set the pattern for later developments. Our primary concern is to establish conditions for the existence

of solutions X of $X^T AX = B$ where A and B are nonsingular symmetric matrices with real coefficients and n and m variables respectively with $n \geqq m$. With this purpose in mind we defined congruence so that two congruent forms have the same essential properties and a "simplest" form under congruence transformations which we called canonical forms. By this means we were able to establish criteria for the congruence of two forms which did not necessitate reduction of the forms to canonical form. These results enabled us to find conditions for the solvability of the equation $X^T AX = B$ and we were able further to show that X and Y are two solutions if and only if there is an automorph U of A such that $X = UY$. A by-product as well as a useful tool was the last theorem and corollary due to Witt.

Chapter II

FORMS WITH p-ADIC COEFFICIENTS

7. The definition and properties of p-adic numbers. It turns out to be the case that the criteria for congruence and representation of numbers by forms with rational coefficients can be reduced to consideration of like problems for forms with p-adic coefficients. As was stated in the introduction, the results of this chapter could be stated largely as Minkowski gave them in terms of congruences. Pall has done just this. But since the gain in their introduction seems to justify such treatment, we prove our results in terms of p-adic numbers.

For p a prime number, we define a *p-adic number* to be any formal series

$$\alpha = a_{-r}p^{-r} + a_{-r+1}p^{-r+1} + \cdots + a_{-1}p^{-1} + a_0 + a_1p + a_2p^2 + \cdots \tag{3}$$

where each a_i is an integer (positive, negative or zero). We call two p-adic numbers α, β *equal* if there is an integer K such that for every positive integer $k \geqq K$, $\alpha_k - \beta_k \equiv 0(\text{mod } p^k)$ where α_k is the integer obtained from the expansion of α by deleting all terms in the expansion of α after that involving p^k and β_k is similarly defined. This definition of equality has the usual characteristic properties of equality: $\alpha = \alpha$; $\alpha = \beta$ implies $\beta = \alpha$; $\alpha = \beta$, $\beta = \gamma$ implies $\alpha = \gamma$.

Any p-adic number may be put into so-called *canonical form,* that is, written in the form (3) above where each a_i is a non-negative integer less than p. This is done as follows. Let $a_{-r} = a'_{-r} + pw$ where $0 \leqq a'_{-r} < p$ and let α' be the p-adic number obtained from α by replacing

$a_{-r}p^{-r} + a_{-r+1}p^{-r+1}$ by $a'_{-r}p^{-r} + (w + a_{-r+1})p^{-r+1}$. It is clear that $\alpha = \alpha'$. This process may be used to replace α by an equal p-adic number in canonical form. Two p-adic numbers in canonical form are equal if and only if they have identical coefficients.

If β is a p-adic number with coefficients b_i, define $\alpha + \beta$ to be the p-adic number with coefficients $a_i + b_i$. In fact, one may add two canonical forms to get a canonical form in a manner entirely analogous to the way in which we add decimals except that we add from left to right instead of from right to left. For example we add the two 5-adic numbers below as follows:

$$\alpha = 3\cdot 5^{-2} + 4\cdot 5^{-1} + 2 + 2\cdot 5 + 2\cdot 5^2 + \cdots$$

$$\beta = 2\cdot 5^{-2} + 0\cdot 5^{-1} + 2 + 2\cdot 5 + 2\cdot 5^2 + \cdots$$

$$\begin{array}{lccccc} \text{Carry} & & 1 & 1 & 1 & 1 \\ \hline \alpha + \beta = & 0\cdot 5^{-2} & + 0\cdot 5^{-1} & + 0 & + 0\cdot 5 & + 0\cdot 5^2 + \cdots \end{array}$$

The product of two p-adic numbers in form (3) is the formal product:

$$a_{-r}b_{-r}p^{-2r} + (a_{-r}b_{-r+1} + b_{-r}a_{-r+1})p^{-2r+1} + (a_{-r}b_{-r+2} + a_{-r+1}b_{-r+1} + a_{-r+2}b_{-r})p^{-2r+2} + \cdots$$

This, again may be reduced to canonical form.

Any p-adic number whose coefficients beyond a certain point are all zero we write as a formal series with a finite number of terms, for omitting such terms does not alter the value of its sum or product with any other p-adic number. Thus $1 + 0\cdot p + 0\cdot p^2 + \cdots$ we write as 1 and notice that when this is multiplied by any p-adic number, the number is unchanged. Similarly $0 = 0 + 0\cdot p + 0\cdot p^2 + \cdots$ has the property that $0 + \alpha = \alpha$ for any p-adic number α. In a similar fashion for any prime p, any integer N may be written as a p-adic integer whose coefficients from a certain point on are all zero. For instance 36 is identified with the 5-adic integer $1 + 2\cdot 5 + 1\cdot 5^2$.

If α is written in the form (3), $-\alpha$ is obtained by changing the sign of all the coefficients and, from the definition of equality, $\alpha + \gamma = 0 = \alpha + \gamma'$ implies $\gamma = \gamma'$, that is, $-\alpha$ is unique.

A p-adic number α is called a *p-adic integer* if it is equal to a p-adic number

$$a_0 + a_1p + a_2p^2 + \cdots .$$

If, in this form, $a_0 \not\equiv 0 (\text{mod } p)$ the number is called a *p-adic unit*. If α is a p-adic unit we can find in the following manner a p-adic unit β such that $\alpha\beta = 1$. Write $\beta_0 = b_0 + b_1p + b_2p^2 + \cdots$ and multiplying α and β we have

$$\alpha\beta = a_0b_0 + (a_0b_1 + a_1b_0)p + (a_0b_2 + a_1b_1 + a_2b_0)p^2 + \cdots . \tag{4}$$

Since a_0 is prime to p we can find a non-negative integer b_0 less than p such that $a_0b_0 \equiv 1 (\text{mod } p)$. (Notice that it is at this point that the requirement that p be a prime is necessary.) Make this choice for b_0 and write $a_0b_0 = 1 + t_0p$. Notice that $b_0 \neq 0$ and let b_1 be that non-negative integer less than p for which $a_0b_1 + a_1b_0 + t_0 \equiv 0 (\text{mod } p)$ and write $a_0b_1 + a_1b_0 + t_0 = t_1p$. Let b_2 be that non-negative integer less than p for which $a_0b_2 + a_1b_1 + a_2b_0 + t_1 = t_2p$. Continuing in this fashion we can develop the formal series for β, the unique p-adic number which is the reciprocal of α. In fact, the reciprocal of every p-adic number γ not zero may be found since γ may be written in the form $p^k\alpha$ where α is a unit and therefore expressible in the form (3) with $r = 0$. Then γ^{-1} may be identified with the number

$p^{-k}\beta$ where β is a unit determined as above. For example, to find the reciprocal of the particular p-adic number α above, let $\alpha' = 5^2\alpha$ where $\alpha' = 3 + 4\cdot 5 + 2\cdot 5^2 + 2\cdot 5^3 \cdots$. Then $a_0b_0 = 3b_0 = 1 + 5t_0$ shows that $b_0 = 2$, $t_0 = 1$; $3b_1 + 4b_0 + t_0 = 3b_1 + 8 + 1 = 5t_1$ shows that $b_1 = 2$, $t_1 = 3$. So continuing we find

$$\beta = 2 + 2\cdot 5 + 0\cdot 5^2 + 3\cdot 5^3 + 2\cdot 5^4 + 4\cdot 5^5 + 1\cdot 5^6 + 2\cdot 5^7 + 0\cdot 5^8 + \cdots$$

where the sequence 2, 0, 3, 2, 4, 1 of coefficients repeats indefinitely. Then the reciprocal of α is $5^2\beta$.

Since $\alpha\xi = 1$ is solvable in p-adic numbers for α a non-zero p-adic number, so also is $\alpha\xi = \beta$ where β is a p-adic number. In particular if α and β are integers, ξ may then be identified with the rational number β/α. We should notice that if a and b are integers and a/b is in lowest terms it is a p-adic integer if b is prime to p and a p-adic unit if a and b are both prime to p. In fact, a p-adic unit may be defined to be a p-adic integer whose reciprocal is a p-adic integer.

We collect all the above results and a few others more easily shown into the statement: for any prime p, the p-adic numbers form a *field*, that is, they satisfy the following requirements:

If α, β and γ are p-adic numbers

1. $\alpha + \beta$ and $\alpha\beta$ are unique p-adic numbers (the closure property).
2. $\alpha + \beta = \beta + \alpha$ and $\alpha\beta = \beta\alpha$ (the commutative property).
3. There exist numbers 0 and 1 such that $\alpha + 0 = \alpha$ and $\alpha\cdot 1 = \alpha$ for every p-adic number α.
4. For every p-adic number $\alpha \neq 0$, there is a p-adic number β such that $\alpha\beta = 1$.

5. For every p-adic number α, there is a p-adic number β such that $\alpha + \beta = 0$.
6. $(\alpha + \beta)\gamma = \alpha\gamma + \beta\gamma$ (the distributive property).
7. $(\alpha + \beta) + \gamma = \alpha + (\beta + \gamma)$ and $(\alpha\beta)\gamma = \alpha(\beta\gamma)$ (the associative property).

We denote the field of p-adic numbers by the symbol $F(p)$. Real numbers form a field which we denote by $F(\infty)$ or $R(\infty)$. The rational numbers form a field which is contained in every p-adic field. But the fourth requirement above fails to hold for integers. Since the other six requirements hold for the set of integers as well as the set of p-adic integers we say that each of these sets form a *ring* and denote by $R(p)$ the ring of p-adic integers. (Actually the requirements for an abstract ring are less stringent in that the commutative property for multiplication is not assumed.)

Using the usual number-theoretic notation we say that $\alpha \equiv \beta \pmod{\gamma}$ for p-adic integers α, β and γ if $(\alpha - \beta)/\gamma$ is a p-adic integer, that is, if $\alpha - \beta$ is divisible by γ. We speak of a form (or matrix) as being "in $R(p)$" or "in $F(p)$" if its coefficients (or elements) are in the ring or field.

8. **Congruences and p-adic numbers.** The connection between p-adic numbers and congruences is shown by the following useful theorem.

THEOREM 9a. *If two quadratic forms*

$$f = \sum_{i,j=1}^{n} \alpha_{ij} x_i x_j , \qquad g = \sum_{i,j=1}^{m} \beta_{ij} y_i y_j$$

have matrices A and B respectively in $R(p)$, if $m \leqq n$ and if T is a matrix with rational integral elements such that $T^T A T \equiv B \pmod{p^{u+w}}$ where p^u is the highest power of p in $|B|$ and $w = 1$ or 3 according as p is odd or $p = 2$, then there is a *matrix in* $R(p)$ *such that* $X^TAX = B$ *and* $X \equiv T \pmod p$ *or* (mod 4) *according as* p *is odd or* $p = 2$.

To prove this by induction let $S_t^TAS_t - B = p^tU$ where $t \geqq u + w$ and S_t and U are in $R(p)$. This holds for $S_t = T$ when $t = u + w$.

Define

$$S_{t+1} = S_t - \tfrac{1}{2}p^tS_t(S_t^TAS_t)^IU.$$

If S_t is in $R(p)$, then S_{t+1} is also for $S_t^TAS_t = B + p^tU$ implies that p^u is the highest power of p in $|S_t^TAS_t|$; furthermore, $t \geqq u + w$ implies $S_{t+1} \equiv S_t \pmod p$ or (mod 4) according as p is odd or $p = 2$. Then

$$\begin{aligned} S_{t+1}^TAS_{t+1} - B \\ &= [S_t - \tfrac{1}{2}p^tU(S_t^TAS_t)^IS_t^T]A[S_t - \tfrac{1}{2}p^tS_t(S_t^TAS_t)^IU] - B \\ &= p^tU - \tfrac{1}{2}p^tU - \tfrac{1}{2}p^tU + \tfrac{1}{4}p^{2t}U(S_t^TAS_t)^IU \\ &= \tfrac{1}{4}p^{2t}U(S_t^TAS_t)^IU \equiv 0 \pmod{p^{2t-u}/4}. \end{aligned}$$

Now if p is odd $2t - u \geqq t + 1$ since $t \geqq u + 1$ and if $p = 2$, $2t - u - 2 \geqq t + 1$ since $t \geqq u + 3$. Hence

$$S_{t+1}^TAS_{t+1} \equiv B \pmod{p^{t+1}}$$

and by this means our theorem is proved, for the successive S_t generate X in $R(p)$.

The following companion theorem is also useful

THEOREM 9b. *If* f *and* g *are defined as in theorem* 9a *and if* T *is a matrix with rational integral elements such that* $T^TAT \equiv B \pmod{p^{2u+1}}$, *not every* m-*rowed minor determinant of* T *is divisible by* p *and* p^u *is the highest power of* p *dividing every* m-*rowed minor of* $2T^TA$; *then there is a matrix* X *in* $R(p)$ *such that* $X^TAX = B$ *and* $X \equiv T \pmod{p^{u+1}}$. *Furthermore* $u \leqq mk$ *where* k *is the highest power of* p *in* $2|A|$.

First we show that $u \leqq mk$. Suppose every m by m minor determinant of $2T^TA = R$ is divisible by p^{mk+1}. Then $T^T = \frac{1}{2}RA^I$ and, since the highest power of p in the denominators of the right side is p^k, we have that every m-rowed minor determinant of T is divisible by p, contrary to hypothesis.

Now, in order to establish an induction choose $t \geqq 2u+1$ and let S_t be a matrix with integral elements such that $S_t \equiv T \pmod{p^{u+1}}$ and $S_t^TAS_t - B = p^tU$ where U is in $R(p)$. Then if $S_{t+1} = S_t + p^{t-u}Y$ for Y later to be chosen we have

$$S_{t+1}^TAS_{t+1} - B \equiv p^tU + p^{t-u}Y^TAS_t + p^{t-u}S_t^TAY \pmod{p^{t+1}}.$$

The right side will be divisible by p^{t+1} if we can choose Y so that $2U + p^{-u}2Y^TAS_t + p^{-u}2S_t^TAY \equiv 0 \pmod{2p}$, that is

$$U + p^{-u}(2S_t^TA)Y \equiv 0 \pmod{2p}.$$

Choose A_0 to be a matrix with integral elements and congruent to $A \pmod{p^{2u+1}}$. Now, since the g.c.d. of the m-rowed minors of $2T^TA$ and hence of $2S_t^TA_0$ is p^u we can, from lemma 6 proved later in section 20 find an $n - m$ by n matrix M with integer elements such that

$$\begin{bmatrix} 2S_t^TA_0 \\ M \end{bmatrix} = N$$

has determinant p^u. Choose an arbitrary $n - m$ by m matrix U_0 with integer elements and let

$$Y = -p^uN^I\begin{bmatrix} U \\ U_0 \end{bmatrix}.$$

This matrix Y has integer elements and makes $U +$

$p^{-u}(2S_t^T A)Y$ vanish. Thus we have shown the existence of an S_{t+1} such that

$$S_{t+1}AS_{t+1} \equiv B \pmod{p^{t+1}}, \qquad S_{t+1} \equiv S_t \pmod{p^{u+1}}.$$

In this fashion we may proceed from $T^TAT \equiv B \pmod{p^{2u+1}}$ to the same congruence (mod p^{2u+2}), hence to the next higher power of p and so forth. By this method we can develop a matrix X in $R(p)$ whose elements have the series development obtained.

We shall have occasion to use the following special cases of this theorem:

COROLLARY 9a. *If* $ax^2 + 2by^2 \equiv N \pmod 8$ *is solvable for* a, b *and* N *odd integers, then* $ax^2 + 2by^2 = N$ *is solvable in 2-adic integers.*

COROLLARY 9b. *If a form* f *in* $R(p)$ *has unit determinant and if* $f \equiv \alpha \pmod{p^w}$ *has a solution with* x_i *a unit for some* i, *where* $w = 1$ *or* 3 *according as* p *is odd or even, then* $f = \alpha$ *is solvable in* $R(p)$.

Theorems 9a and 9b can be somewhat loosely described by saying that $T^TAT \equiv B \pmod{p^t}$ for a sufficiently large value of t implies that $X^TAX = B$ is solvable in p-adic integers; the minimum value of t depending on the determinant of B or the g.c.d. of the minors of $2AT$ in the respective cases. This statement is not quite accurate since we must not only have the congruence solvable but there must be a process, an algorithm, by which the solutions in $R(p)$ are developed.

On the other hand, if $X^TAX = B$ has a solution X where X is in $R(p)$, we can in the expansion of each element of X delete the terms from p^t onward and have a solution of the congruence $X^TAX \equiv B \pmod{p^t}$. This result we state as

THEOREM 10. *If* f *and* g *are forms with matrices* A *and* B

having integer elements and if $X^TAX = B$ *has a solution in* $R(p)$, *then for any positive integer* t, *the congruence* $X^TAX \equiv B \pmod{p^t}$ *has a solution* T, *where* T *is a matrix with integer elements for which* $T \equiv X \pmod{p^t}$.

Theorems 9 and 10 establish a $1 - 1$ correspondence between the solutions of $X^TAX = B$ in $R(p)$ and the solutions of $X^TAX \equiv B \pmod{p^t}$ for t sufficiently large.

As an example of the application of these three theorems let f be the quadratic form $5x^2$ and consider the solutions of $5x^2 = 2$ in 3-adic, 5-adic and 2-adic integers. If $p = 3$, the theorems show that the equation is solvable in $R(3)$ if and only if $5x^2 \equiv 2 \pmod 3$ is solvable. Now this congruence has the solution $x = 1$ and $x = 2$. The method of proof of theorem 9b yields a solution as follows. Write $x = 1 + 3x_1$ and seek an x_1 so that $5(1 + 3x_1)^2 \equiv 5(1 + 6x_1) \equiv 2 \pmod 9$, i.e. $x_1 = 2 + 3x_2$ or $x = 7 + 9x_2$. Thus we wish to choose an x_2 so that $5(7 + 9x_2)^2 \equiv 2 \pmod{81}$. So proceeding we get

$$1 + 2\cdot 3 + 0\cdot 3^2 + 0\cdot 3^3 + 0\cdot 3^4 + 2\cdot 3^5 + 1\cdot 3^6 + \cdots$$

$$2 + 0\cdot 3 + 2\cdot 3^2 + 2\cdot 3^3 + 2\cdot 3^4 + 0\cdot 3^5 + 1\cdot 3^6 + \cdots$$

as the two solutions in $R(3)$. (The second may be obtained from the first by noting that their sum is zero.) Similarly $5x^2 = 2$ is solvable in $R(5)$ if and only if $5x^2 \equiv 2 \pmod 5$ is solvable; neither the congruence nor the equation have solutions. Finally the insolubility of the congruence $5x^2 \equiv 2 \pmod 8$ shows that the equation is not solvable in $R(2)$.

9. **Congruence of forms in $F(p)$.** In a manner analogous to that in the previous chapter we say that two forms f and g in $F(p)$ are *congruent in* $F(p)$ if there is a non-singular transformation with p-adic elements taking f into g.

Theorem 1 for the field $F(p)$ states that every form f with p-adic coefficients may be taken into a form

$$g = \beta_1 y_1^2 + \beta_2 y_2^2 + \cdots + \beta_n y_n^2, \quad \beta_i \text{ in } F(p)$$

by a linear transformation whose elements are p-adic numbers.

It is possible to simplify further this canonical form for p-adic numbers. To this end notice that if β is any p-adic number, there is an integer t such that $p^{2t}\beta = \beta_0$ is a p-adic integer not divisible by p^2. Then if b is the integer obtained by deleting all after the term involving p^5 in the expansion of β_0 we have $\beta_0 \equiv b \pmod{p^5}$ and hence, by theorem 9a, there is a p-adic integer α such that $\alpha^2\beta_0 = b \not\equiv 0 \pmod{p^2}$. Hence for any p-adic number β_i there is a p-adic integer α_i such that $\alpha_i^2\beta_i = b_i$, an integer not divisible by p^2. Thus, using the transformation $y_i = \alpha_i x_i$ on g we have

THEOREM 11. *Every form f in $F(p)$ is congruent in $F(p)$ to a form*

$$g = b_1 x_1^2 + b_2 x_2^2 + \cdots + b_n x_n^2$$

where the b_i are integers not divisible by p^2.

It is possible to carry the process further and obtain a unique canonical form but this would be rather laborious especially for $p = 2$. The main purpose in having a unique form, namely testing two forms for congruence, is better served for $F(p)$ by means of invariants defined in the next two sections.

10. **The Hilbert symbol.** Hilbert defined a symbol which we shall find very useful in slightly modified form. For α and β non-zero p-adic numbers we define the symbol

$$(\alpha, \beta)_p = +1 \text{ or } -1$$

according as

$$\alpha x_1^2 + \beta x_2^2 = 1$$

has or has not a solution in $F(p)$. Recall that we include as values of p not only the prime numbers but the "infinite prime" and that $F(\infty)$ and $R(\infty)$ both denote the field of real numbers. Where no ambiguity results we often omit the subscript on the parentheses.

For easy reference let us first list the properties of the Hilbert symbol which we shall find useful and then give the proofs of those properties which do not directly follow from the definition. Unless otherwise stated each property holds for $p = \infty$ as well as for p finite, but the statement "p odd" of course restricts p to be finite. If α is a unit in $F(p)$ for p finite we understand $(\alpha \mid p)$ to be the value of the Legendre symbol $(a_0 \mid p)$ where a_0 is the leading term in the p-adic expansion of α. Below it is understood that $\alpha, \beta, \gamma, \rho, \sigma$ are non-zero numbers in $F(p)$.

1. $(\alpha, \beta)_\infty = 1$ unless α and β are both negative.
2. $(\alpha, \beta)_p = (\beta, \alpha)_p$.
3. $(\alpha\rho^2, \beta\sigma^2)_p = (\alpha, \beta)_p$.
4. $(\alpha, -\alpha)_p = 1$.
5. If $\alpha = p^a\alpha_1$, $\beta = p^b\beta_1$ with α_1 and β_1 units, then
   a. if p is odd, $(\alpha, \beta)_p = (-1 \mid p)^{ab}(\alpha_1 \mid p)^b(\beta_1 \mid p)^a$
   b. if $p = 2$, $(\alpha, \beta)_2$
   $$= (2 \mid \alpha_1)^b(2 \mid \beta_1)^a(-1)^{(\alpha_1-1)(\beta_1-1)/4}.$$

5′. If p is prime to $2\alpha\beta$, $(\alpha, \beta)_p = 1$, for p finite, α and β in $R(p)$.

6. $(\alpha, \beta)_p(\alpha, \gamma)_p = (\alpha, \beta\gamma)_p$.
7. $(\alpha, \alpha)_p = (\alpha, -1)_p$.
8. $(\alpha\rho, \beta\rho)_p = (\alpha, \beta)_p(\rho, -\alpha\beta)_p$.
9. If β is a non-square in $F(p)$ and $c = 1$ or -1, there is for each prime p an integer α such that $(\alpha, \beta)_p = c$.

If, further, b as defined in property 5 is odd, α may be taken prime to p.

10. If a and b are non-zero rational numbers.
$$\prod (a, b)_p = 1$$
the product extending over all primes p including $p = \infty$.

Properties 1, 2 and 3 are obvious from the definition of the symbol. Property 4 holds since $\alpha(x_1^2 - x_2^2) = 1$ has a solution $x_1 = (1 + \alpha^{-1})/2$, $x_2 = (1 - \alpha^{-1})/2$.

To prove property 5 notice that property 3 shows that we need only consider a, b to be 0 or 1. Suppose (ξ_0, η_0) is a solution of $\alpha x_1^2 + \beta x_2^2 = 1$ in $F(p)$. We can, from the definition of p-adic numbers, choose a positive integer p^t such that both $\xi_0 p^t$ and $\eta_0 p^t$ are p-adic integers and one is a unit. Thus $(\alpha, \beta)_p = 1$ if and only if for some positive integer t, $\alpha x_1^2 + \beta x_2^2 = p^{2t}$ has a solution (ξ_0', η_0') where ξ_0' and η_0' are p-adic integers and one is a unit. Thus, using theorems 9a and 10, we see that $(\alpha, \beta)_p = 1$ if and only if, for some positive integer t,

$$\alpha x_1^2 + \beta x_2^2 \equiv p^{2t} \pmod{p^{2t+w}} \tag{5}$$

has an integral solution (r_1, r_2) with $w = 1$ or 3 according as p is odd or $p = 2$ and one of r_1 and r_2 is prime to p.

First suppose $a = b = 0$. If $p = 2$, see that (5) has a solution with $t = 0$ unless $\alpha \equiv \beta \equiv 3 \pmod 4$ in which case (5) has no solution with $t = 0$ while $t > 0$ would imply r_1 and r_2 both even, contrary to our assumption. If p is odd, αx_1^2 and $1 - \beta x_2^2$ each take on $(p + 1)/2$ distinct values (mod p) as x_1 and x_2 range over the integers $0, 1, \cdots, (p - 1)/2$. Since $(p + 1)/2 + (p + 1)/2 = p + 1$ there must be some value r_1 of x_1 and r_2 of x_2 such that $\alpha r_1^2 \equiv 1 - \beta r_2^2 \pmod p$ which shows that (5) is solvable for $t = 0$. Thus for both these cases we have established property 5.

Second, suppose $a = 1$, $b = 0$. Since one of r_1, r_2 is a unit we must have $t = 0$. If $p = 2$, (5) is solvable if and only if $\beta = \beta_1 \equiv 1 \pmod 8$ or $\alpha + \beta = 2\alpha_1 + \beta_1 \equiv 1 \pmod 8$ in which cases

$$(2 \mid \beta_1) = 1 = (-1)^{(\alpha_1-1)(\beta_1-1)/4} \text{ or}$$

$$(2 \mid \beta_1) = (-1)^{(\alpha_1-1)/2} \text{ with } \beta_1 \equiv 3 \pmod 4$$

which establishes the result. If p is odd (5) is solvable if and only if $(\beta \mid p) = 1$.

Finally if $a = 1 = b$, the equation $\alpha x_1^2 + \beta x_2^2 = p^{2t}$ becomes $\alpha_1 x_1^2 + \beta_1 x_2^2 = p^{2t-1}$ and in place of (5) we have

$$(5') \qquad \alpha_1 x_1^2 + \beta_1 x_2^2 \equiv p^{2t-1} \pmod{p^{2t+w-1}}.$$

If $p \neq 2$ suppose $(-\alpha_1\beta_1 \mid p) = -1$; then (5′) implies $x_1 = p^{t-1}x_1'$, $x_2 = p^{t-1}x_2'$ and $\alpha_1 x_1'^2 + \beta_1 x_2'^2 \equiv p \pmod{p^2}$ which implies x_1' and x_2' prime to p and $(-\alpha_1\beta_1 \mid p) = 1$ which is a contradiction. On the other hand, the last congruence is solvable if $(-\alpha_1\beta_1 \mid p) = 1$; thus for $p \neq 2$ equation (5′) is solvable if and only if $(-\alpha_1\beta_1 \mid p) = 1$. If $p = 2$, (5′) is solvable with x_1x_2 odd if and only if $t = 1$ with $\alpha_1 + \beta_1 \equiv 2 \pmod 8$ or $t > 1$ with $\alpha_1 + \beta_1 \equiv 0 \pmod 8$. Let $\alpha_1 + \beta_1 \equiv 2k \pmod 8$. Then $1 + \alpha_1\beta_1 \equiv 2k\alpha_1 \pmod 8$, $(\alpha_1 - 1)(\beta_1 - 1) = \alpha_1\beta_1 + 1 - \alpha_1 - \beta_1 \equiv 2k\alpha_1 - 2k \pmod 8$ and hence

$$(2 \mid \alpha_1)\,(2 \mid \beta_1)(-1)^{(\alpha_1-1)(\beta_1-1)/4}$$
$$= (2 \mid 2k\alpha_1 - 1)(-1)^{k(\alpha_1-1)/2}.$$

This is $+1$ if $k = 0$ or 1 and -1 if $k = 2$ or 3 which shows that $(\alpha, \beta)_2 = 1$ if and only if $\alpha_1 x_1^2 + \beta_1 x_2^2 \equiv 2^{2t-1} \pmod 8$ has a solution with x_1x_2 odd and $t = 1$ or 2. But corollary

9b shows that under these circumstances (5′) is solvable and our proof of property 5 is complete.

Property 5′ follows directly from property 5.

Property 6 for $p = \infty$ follows directly from the definition. If p is odd, property 5 tells us that

$$(\alpha, \beta)_p(\alpha, \gamma)_p = (-1 \mid p)^{a(b+c)}(\alpha_1 \mid p)^{(b+c)}(\beta_1\gamma_1 \mid p)^a$$

which implies property 6 since $\beta\gamma = \beta_1\gamma_1 p^{b+c}$. If $p = 2$, property 5 yields

$$(\alpha, \beta)_2(\alpha, \gamma)_2 = (2 \mid \alpha_1)^{b+c}(2 \mid \beta_1\gamma_1)^a(-1)^t$$

with $t = (\alpha_1 - 1)(\beta_1 - 1 + \gamma_1 - 1)/4 \equiv (\alpha_1 - 1)(\beta_1\gamma_1 - 1)/4 \pmod 2$.

Property 7 follows from properties 4 and 6; property 8 from properties 2, 6 and 7. To prove property 9 with $c = 1$, choose α a square; if $c = -1$ for $p = \infty$ choose $\alpha = -1$, for p odd choose $\alpha = p$ or a non-residue of p according as b is even or odd, for $p = 2$ and b odd take $\alpha = \alpha_1$ odd so that

$$(2 \mid \alpha_1)(-1)^{(\alpha_1-1)(\beta_1-1)/4} = -1,$$

for $p = 2$ and b even choose $\alpha = 2$ and see that $(\alpha, \beta)_2 = (2 \mid \beta_1) = -1$ unless $\beta_1 \equiv 7 \pmod 8$ when we choose $\alpha = 6$.

Finally, to prove property 10 we see from property 3 that α and β may be restricted to be p-adic integers and hence, by property 5, to be rational square-free integers. Hence we may write $|\alpha| = q_1q_2 \cdots q_t$, $|\beta| = r_1r_2 \cdots r_t$ where the q's and r's are primes, no two q's equal and no two r's equal. If either α or β is ± 1 let it be α. Thus, by property 6 we have

$$(\alpha, \beta)_p = \begin{cases} \prod_{i,j}(q_i, r_j)_p \text{ if neither } \alpha \text{ nor } \beta \text{ is } \pm 1, \\ \prod_j(\pm 1, r_j)_p \text{ if } \alpha = \pm 1, \end{cases}$$

where in the second product r_j is replaced by b if $b = 1$ or -1, the sign of α is included in q_1 and that of β in r_1. If either α or β is $+1$, the symbol has the value 1. Hence we need prove property 10 only for the following values of $(\alpha, \beta)_p$,

$$(-1, -1)_p, \quad (-1, r)_p, \quad (q, q)_p, \quad (q, r)_p \text{ with } q \neq r$$

and q and r positive primes. The third reduces to the second by property 7. The remainder of the proof follows taking r and q odd primes:

$$\prod_p (-1, -1)_p = (-1, -1)_2(-1, -1)_\infty = 1$$

$$\prod_p (-1, 2)_p = (-1, 2)_2 = 1$$

$$\prod_p (-1, r)_p = (-1, r)_r(-1, r)_2 = 1$$

$$\prod_p (2, q)_p = (2, q)_2(2, q)_q = 1$$

$$\prod_p (r, q)_p = (r, q)_q(r, q)_r(r, q)_2$$

$$= (q \mid r)(r \mid q)(-1)^{(q-1)(r-1)/4}$$

and the last is equal to 1 by the quadratic reciprocity law.

11. **Hasse's[4] symbol** $c_p(f)$. Let f be a regular form with coefficients in $F(p)$ and non-zero determinant (see section 3) and, using Pall's generalization of Hasse's symbol, we define

$$c_p(f) = (-1, -D_n)_p \prod_{i=1}^{n-1} (D_i, -D_{i+1})_p$$

where the D_i are defined as in section 3, it being understood that if $D_k = 0$, the symbols $(D_k, -D_{k+1})_p$ and $(D_{k-1}, -D_k)_p$ are interpreted to be $(\pm 1, -D_{k+1})_p$ and

$(D_{k-1}, \mp 1)_p$ respectively with the $\pm$ sign chosen arbitrarily. (Note that lemmas 1 and 2 show that f may be taken to be regular and that $D_k = 0$ implies $D_{k+1}D_{k-1} \neq 0$.)

The importance of this symbol is shown by

THEOREM 12. *If f_1 and f_2 are forms in $F(p)$ of non-zero determinants δ_1 and δ_2, having n_1 and n_2 variables respectively and congruent in $F(p)$, p finite or $p = \infty$, then*

$$(6) \qquad \delta_1 = \tau^2\delta_2, \qquad n_1 = n_2, \qquad c_p(f_1) = c_p(f_2)$$

for some element τ of $F(p)$.

Since $f_1 \sim f_2$ implies that the number of variables in each is the same and $\delta_2 = |T|^2 \delta_1$ where T is the transformation taking f_1 into f_2 we see that the burden of the proof is to prove that the Hasse symbols of the two forms are equal.

First we show that there is a diagonal form $f_0 = \sum_{i=1}^{n} \alpha_i x_i^2$ congruent to f and having $c_p(f_0) = c_p(f)$, $\alpha_1 = D_1$ or $\alpha_1\alpha_2 = D_2$ according as $D_1 \neq 0$ or $D_1 = 0$, where D_i is the leading i by i minor of f. If $D_1 \neq 0$ we can add appropriate multiples of the first column of A, the matrix of f, to the other columns and the same for the rows to take f into $D_1x_1^2 + f_1$ with $f_1 = \sum_{i,j=2}^{n} \alpha_{ij}x_ix_j$. This will not alter any principal minor determinant of f and hence does not alter $c_p(f)$. If $D_1 = 0$, then by the regularity of f, $D_2 \neq 0$ and we may take f into $\alpha_{11}x_1^2 + 2\alpha_{12}x_1x_2 + \alpha_{22}x_2^2 + f_1$ with $f_1 = \sum_{i,j=3}^{n}\alpha_{ij}x_ix_j$ without altering any D_i. Now $D_2 \neq 0$ implies $\alpha_{12} \neq 0$ and the transformation

$$\begin{bmatrix} 1 - \alpha_{22}/2\alpha_{12} & -\alpha_{22}/2\alpha_{12} \\ 1 & 1 \end{bmatrix}$$

takes $f_{10} = \alpha_{11}x_1^2 + 2\alpha_{12}x_1x_2 + \alpha_{22}x_2^2$ into $f_{20} = 2\alpha_{12}x_1^2 +$

$2\alpha_{12}x_1x_2 + \alpha_{22}'^2 x_2^2$ with $c_p(f_{10}) = (-1, -D_2)_p(D_1, -D_2)_p$ and $c_p(f_{20}) = (-1, -D_2)_p(2\alpha_{12}, -D_2)_p$. But $-D_2 = \alpha_{12}^2$ and hence both symbols are $+1$ regardless of the choice of D_1. This process may be carried through to produce our result except that we must provide for the following typical contingency. Suppose $D_1 \neq 0$ but the leading coefficient of f_1 is 0. Then our transformation used above to make the leading coefficient of f_1 not zero alters the leading 2 by 2 minor of f, that is $(D_1, -0)_p(0, -D_3)_p$ is replaced by $(D_1, - 2D_1\alpha_{12})_p(+2D_1\alpha_{12}, -D_3)_p$. Both products are $(-1, -D_3)_p$ since $-D_1D_3$ is a square.

Second, for a diagonal form f_0, $c_p(f_0)$ remains unchanged under any permutation of subscripts. To show this it is sufficient to show that $c_p(f_0) = c_p(f_0')$ where f_0' is obtained from f_0 by interchanging the r-th and $(r - 1)$st subscripts. Such an interchange leaves unaltered every D_i except D_{r-1} which becomes D_{r-1}', the minor obtained by interchanging the r-th and $(r - 1)$th rows and columns of D_r' and then deleting the rth row and column. Thus we need to show

$$(D_{r-1}, -D_r)(D_{r-2}, -D_{r-1}) = (D_{r-1}', -D_r)(D_{r-2}, -D_{r-1}').$$

But, using the properties of the Hilbert symbol, we see that the left hand side of the equation is equal to $(D_{r-1}, -D_rD_{r-2})(D_{r-2}, -1)$. Thus we need to show that $(D_{r-1}, -D_rD_{r-2}) = (D_{r-1}', -D_rD_{r-2})$, that is

$$(D_{r-1}D_{r-1}', -D_rD_{r-2}) = 1. \tag{7}$$

Now lemma 1 tells us that $D_rD_{r-2} = D_{r-1}D_{r-1}' - D_{r,r-1}^2$ for some determinant $D_{r,r-1}$. But $D_{r-1}D_{r-1}'x_1^2 + (D_{r,r-1}^2 - D_{r-1}D_{r-1}')x_2^2 = 1$ has the solution $x_1 = x_2 = D_{r,r-1}^{-1}$ and hence (7) holds if $D_{r,r-1} \neq 0$ while if $D_{r,r-1} = 0$, property 4 of the Hilbert symbol shows (7) holds. The argument of this paragraph need not have been confined to diagonal forms except that we need to be assured that no D_i is zero. This assurance follows for diagonal forms from $|f| \neq 0$ but does not follow for forms in general.

We have reduced our problem to showing that

$$f = \sum_{i=1}^{n} \alpha_i x_i^2 \backsim g = \sum_{i=1}^{n} \beta_i y_i^2$$

and $|f| \cdot |g| \neq 0$ implies $c_p(f) = c_p(g)$. The first column of a transformation taking f into g is a solution $(\rho_1, \rho_2, \cdots, \rho_n)$ of $f = \beta_1$. Permuting the variables of f if necessary we assume $\rho_1 \neq 0$. Then the transformation T whose first row is $\rho_1, 0, \cdots, 0$, whose first column is $\rho_1, \rho_2, \cdots, \rho_n$ and the rest of which is the identity matrix with $n - 1$ rows, takes f into a form f' whose leading coefficient is β_1. Now $T = T_1 T_2 \cdots T_n$ where T_i is the matrix obtained from the identity transformation by replacing the i-th element in its first column by ρ_i. It is clear that T_1 replaces every D_i by $\rho_1^2 D_i$ and hence does not alter the Hasse invariant of f. The transformation T_2 alters only D_1 which it replaces by $\alpha_1 + \rho_2^2\alpha_2$ and $(\alpha_1, -\alpha_1\alpha_2)_p(\alpha_1 + \rho_2^2\alpha_2, -\alpha_1\alpha_2)_p = (\alpha_1^2 + \rho_2^2\alpha_1\alpha_2, -\alpha_1\alpha_2)_p = 1$ since $(\alpha_1^2 + \rho_2^2\alpha_1\alpha_2)x_1^2 - \alpha_1\alpha_2 x_2^2 = 1$ has the solution $x_1 = \alpha_1^{-1}$, $x_2 = \rho_1\alpha_1^{-1}$. Thus T_2 does not alter the Hasse invariant. Now T_i, $i > 2$, is obtained from T_2 by a permutation of rows and columns which is equivalent to a permutation of the variables of the form and the application of T_2. Neither of these alters the Hasse invariant. Hence the transformation T leaves unaltered the Hasse invariant. By the process of the first portion of this proof we may take f' into $\beta_1 x_1^2 + f_1$ with $f_1 = \sum_{i,j=2}^{n} \alpha'_{ij} x_i x_j$ without altering the Hasse invariant. By theorem 8, $f \sim g$ implies $f_1 \sim g_1$ where g_1 is obtained from g by deleting the first term. Hence we may assume by induction that $c_p(f_1) = c_p(g_1)$.

Finally, by the definition of the Hasse symbol, $c_p(\beta_1 x_1^2 + f_1)$ is

$$(-1, -\beta_1 \mid f_1 \mid)_p(\beta_1, -\beta_1 D_1')_p(\beta_1 D_1', -\beta_1 D_2')_p \cdots (\beta_1 D_{n-2}', -\beta_1 D_{n-1}')_p$$

where the D_i' are the leading minor determinants of f_1. By property 8 of the Hilbert symbol, this is equal to

$$\begin{aligned}(-1, -D_{n-1}')(1, -D_1')(D_1', -D_2') \cdots (D_{n-2}', -D_{n-1}')\\ \cdot(-1, \beta_1)(\beta_1, D_1')(\beta_1, D_1'D_2') \cdots (\beta_1, D_{n-2}'D_{n-1}')\\ = c_p(f_1)(\beta_1, -D_{n-1}') = c_p(f_1)(\beta_1, \mid f \mid),\end{aligned}$$

omitting the subscripts p on the Hilbert symbols. This shows that $c_p(g) = c_p(f)$ and our proof is complete.

The next few sections enable us to prove the converse of this theorem (see theorem 15) for p finite.

However first let us see what $c_p(f)$ amounts to for

$$f = \alpha_1 x_1^2 + \alpha_2 x_2^2 + \alpha_3 x_3^2$$

where the coefficients are p-adic integers. Then

$$\begin{aligned}c_p(f) &= (-1, -\delta)_p(\alpha_1, -\alpha_1\alpha_2)_p(\alpha_1\alpha_2, -\delta)_p\\ &= (-\alpha_1\alpha_2, -\alpha_2\alpha_3)_p\end{aligned}$$

where δ is the determinant of f. If $p = \infty$, $c_p(f) = 1$ unless $\alpha_1, \alpha_2, \alpha_3$ are all of the same sign. If p is an odd prime not dividing δ, $c_p(f) = 1$ from property 5′ of the Hilbert symbol. We may without loss of generality assume no two α_i have a factor in common and that no coefficient has a square factor; for property 3 of the Hilbert symbol shows that if p^2 divides $\alpha_2\alpha_3$, then $c_p(f) =$

$(-\alpha_1\alpha_2, -\alpha_2\alpha_3/p^2)_p$. Then if p divides α_3 but not $\alpha_1\alpha_2$ and if p^2 does not divide α_3, we have

1. If p is odd, $c_p(f) = (-\alpha_1\alpha_2 \mid p)$.
2. If $p = 2$ and δ is a unit, $c_2(f) = (-1)^t$, where

$$t = (-\alpha_1\alpha_2 - 1)(-\alpha_2\alpha_3 - 1)/4$$
$$= (\alpha_1\alpha_2 + \alpha_1\alpha_3 + \alpha_2\alpha_3 + 1)/4$$
$$\equiv (\alpha_1 + \alpha_2 + \alpha_3 + \delta)/4 \pmod 2.$$

3. If $p = 2$ and $\delta \equiv 2 \pmod 4$, $c_2(f) = (2 \mid -\alpha_1\alpha_2)(-1)^t$ with $t = (-\alpha_1\alpha_2 - 1)(-1 - \alpha_2\alpha_3/2)/4$ and hence $c_2(f) = 1$ if $\alpha_1\alpha_2 \equiv 7 \pmod 8$ or $\alpha_1\alpha_2 \equiv 1 \pmod 4$ with $\alpha_2\alpha_3 \equiv \alpha_1\alpha_2 + 5 \pmod 8$; otherwise $c_2(f) = -1$.

12. Some properties of $c_p(f)$ and $k_p(f)$. We now prove some useful properties of Hasse's symbol for a form f of determinant δ different from zero and having p-adic coefficients. In no case but the first is p restricted to be finite. The proofs follow the listing of the properties.

1. If f has coefficients in $R(p)$, where p is a finite prime not dividing 2δ, then $c_p(f) = 1$.
2. If f has rational coefficients, then for no primes p or an even number of primes p (including $p = \infty$) is $c_p(f) = -1$.
3. If ω is a number of $F(p)$,

$$c_p(\omega f) = (\omega, \{-1\}^{(n+1)/2})_p c_p(f) \text{ if } n \text{ is odd}$$
$$c_p(\omega f) = (\omega, \{-1\}^{n/2}\delta)_p c_p(f) \text{ if } n \text{ is even.}$$

4. If f_1 and f_2 are two forms,

$$c_p(f_1 + f_2) = c_p(f_1)c_p(f_2)(-1, -1)_p(\delta_1, \delta_2)_p$$

where f_1 and f_2 have no variables in common and have determinants δ_1 and δ_2. A special case of this property is

4′. $c_p(f_1 + \alpha x^2) = c_p(f_1)(-1, \alpha)_p(\delta_1, \alpha)_p$
$$= c_p(f_1)(-\delta_1, \alpha)_p = c_p(f_1)(\delta, \alpha)_p .$$

5. If we follow Pall and define for an n-ary form f with n odd the invariant
$$k_p(f) = c_p(f)(\delta, \{-1\}^{(n+1)/2})_p$$
it has the property
$$k_p(\omega f) = k_p(f),$$
for any non zero number ω of $F(p)$.

Considering property 1 notice that the restriction that the coefficients of f be p-adic integers is necessary since, for instance, $c_5(5x_1^2 + 2x_2^2/5) = -1$. Property 1 follows immediately from the properties of the Hilbert symbol if all D_i are prime to p; this includes the case f diagonal. But, by theorem 1, every form is congruent to a diagonal form g and, by theorem 12, $c_p(f) = c_p(g)$.

The second property above follows directly from property 10 of the Hilbert symbol.

To prove property 3 write

$$c_p(\omega f) = (-1, -\omega^n \delta)_p \prod_{i=1}^{n-1} (\omega^i D_i, -\omega^{i+1} D_{i+1})_p$$
$$= (-1, -\omega^n \delta)_p$$
$$\cdot \prod_{i \text{ even}}^{n-1} (D_i, -\omega D_{i+1})_p \prod_{i \text{ odd}}^{n-1} (\omega D_i, -D_{i+1})_p$$
$$= (-1, -\omega^n \delta)_p \prod_{i=1}^{n-1} (D_i, -D_{i+1})_p T$$
$$= (-1, \omega^n)_p c_p(f) T,$$

where

$$T = \prod_{i \text{ even}}^{n-1} (\omega, D_i)_p \prod_{i \text{ odd}}^{n-1} (\omega, -D_{i+1})_1$$
$$= (\omega, (-1)^t \prod_{i \text{ even}} D_i \prod_{i \text{ odd}} D_{i+1})_p ,$$

and $t = (n - 1)/2$ or $n/2$ according as n is odd or even, since the number of negative signs in the product is equal to the number of odd numbers not greater than $n - 1$. But the product in the last equation is λ^2 or $\lambda^2\delta$ for some λ in $F(p)$ according as n is odd or even. Thus if n is odd $c_p(\omega f) = (\omega, -1)_p(\omega, \{-1\}^t)_p c_p(f)$, while if n is even, $c_p(\omega f) = (\omega, \{-1\}^t\delta)_p c_p(f)$. This completes the proof.

To prove property 4 use the same argument used in the final part of the proof of theorem 12. Property 5 follows from its definition and property 3.

13. **Evaluation of the Hasse symbol.** Theorem 11 shows that every form in $F(p)$ is congruent in $F(p)$ to

$$f = f_1 + pf_2$$

where f_1 and f_2 are forms with integer coefficients and determinants prime to p. By properties 3 and 4 of the Hasse symbol $c_p(f)$ can be evaluated in terms of $c_p(f_1)$ and $c_p(f_2)$ and the determinants of f_1 and f_2. Then if p is odd we have, using property 1 of the Hasse symbol, a complete evaluation.

It remains then to evaluate $c_p(f)$ for $p = \infty$ and $p = 2$ with $|f|$ odd. A glance at property 5 of the Hilbert symbol shows that the value of $c_2(f)$ for $|f|$ odd depends only on the value of the coefficients (mod 4). Thus if f is taken to be diagonal (by theorem 11) and if i of its coefficients are congruent to 1 (mod 4), $c_2(f) = c_2(g)$ where

$$g = x_1^2 + x_2^2 + \cdots + x_i^2 - x_{i+1}^2 - \cdots - x_n^2 .$$

Now, properties 1 and 2 of the Hasse symbol show us that $c_2(g) = c_\infty(g)$. Hence we need merely to find $c_\infty(g)$. In order to make use of properties 3 and 4 of the Hasse symbol write $g = g_1 - g_2$, where each g_i is the sum of squares and have

$$c_\infty(g) = c_\infty(g_1)c_\infty(-g_2)(-1, -1)_\infty(1, \{-1\}^{n-i})_\infty$$
$$= (-1)(-1, u)_\infty(-1)(-1),$$

where $u = (-1)^{(n-i+1)/2}$ or $(-1)^{3(n-i)/2}$ according as $n - i$ is odd or even. This reduces to

$$(8) \qquad c_2(g) = c_\infty(g) = (-1)^{(n-i-2)(n-i-1)/2},$$

that is, $c_2(g) = c_\infty(g) = 1$ if $n - i \equiv 1$ or $2 \pmod 4$ and otherwise -1. Thus $c_2(f)$ and $c_\infty(f)$ may be evaluated by (8) where for the former i is the number of coefficients congruent to 1 (mod 4) in the diagonal form congruent to f in $F(2)$ and for the latter i is the index of the form.

Now we saw in section 3 that it was not necessary to convert a form into canonical form in order to find its index. In just the same way it may be shown that any form with unit determinant in $F(2)$ is congruent in $F(2)$ to a form in which no two successive D_i are non-units. The analogy for determination of the index breaks down if one of the D's is congruent to 2 (mod 4) but if it happens that each D is congruent to 0, 1 or -1 (mod 4) and no two successive D are divisible by 4, then the number i above for $F(2)$ is the number of permanencies of sign in the sequence

$$1\ D_1\ D_2\ \cdots\ D_n \pmod 4$$

where any $D_i \equiv 0 \pmod 4$ may be given arbitrary sign. Thus if

$$f_1 = 2x_1x_2 + 2x_3x_4 + \cdots + 2x_{2n-1}x_{2n},$$

the number of permanences of sign in the sequence of D_i (mod 4) is n and hence

$$c_2(f_1) = (-1)^{(n-2)(n-1)/2}.$$

On the other hand, $c_2(f_2)$ cannot be so evaluated for

$$f_2 = 2x_1x_2 + \cdots + 2x_{2n-3}x_{2n-2} + 2x_{2n-1}^2 + 2x_{2n-1}x_{2n} + 2x_{2n}^2.$$

However, the application of the above result and property 4 of the Hasse symbol yields

$$c_2(f_2) = -c_2(f_1).$$

Notice that $c_\infty(f)$ determines the index of f only to within a multiple of 4. For instance

$$f = -x_1^2 - x_2^2 - x_3^2 - x_4^2 + x_5^2$$

$$f' = +x_1^2 + x_2^2 + x_3^2 + x_4^2 + x_5^2$$

have the same values for $c_\infty(f)$ but their indices are not equal. Thus the converse of theorem 12 does not hold for $p = \infty$. However notice that if i is the index of the form it must satisfy the conditions

(9a) $d(-1)^{n-i} > 0, \quad c_\infty(f) = (-1)^{(n-i-1)(n-i-2)/2}.$

It may be seen that these two conditions may be expressed as one in the form

(9b) $n - i \equiv c_\infty(f) + \frac{1}{2}\{1 + (-1, d)_\infty\} \pmod 4.$

The value of i is determined (mod 4) by $c_\infty(f)$, n and the sign of d.

14. **Zero forms.** Every form with zero determinant represents zero non-trivially (that is, with not all the variables zero) since if the form is canonical some of the coefficients will be zero and the existence of a non-trivial solution is obvious; moreover if a transformation T takes f into canonical form it takes a non-trivial solution into a non-trivial solution. However, there are forms of non-zero determinant which represent zero non-trivially. Such forms are called *zero forms*. One such form is $x_1^2 - x_2^2$.

Notice that a form may be a zero form in one field without being a zero form in another. For instance $x_1^2 + 2x_2^2$ is a 3-adic zero form but not in the field of reals.

If $f = \nu$, with $\nu \neq 0$, then $f - \nu x_{n+1}^2$ is a zero form. Conversely, if $f - \nu x_{n+1}^2 = 0$ has a solution *with* $x_{n+1} \neq 0$, where the coefficients of f and the solution are in some field F, then $f = \nu$ has a solution in F. That the restriction in italics may be removed is shown by the following theorem.

THEOREM 13. *Let the quadratic form f be a zero form in a field F with 2 ± 0 and ν any number in F. Then $f = \nu$ has a solution in F.*

We give Siegel's[14] elegant proof. By theorem 11 we may take

$$f = a_1x_1^2 + a_2x_2^2 + \cdots + a_nx_n^2$$

and call $r_1, r_2, \cdots, r_n$ a solution of $f = 0$ with $r_1 \neq 0$. Set $\tau = \nu(4a_1r_1^2)^{-1}$ and see that $x_1 = (1 + \tau)r_1$, $x_k = (1 - \tau)r_k$, $k = 2, \cdots, n$ is a solution of $f = 4\nu a_1 r_1^2 = \nu$.

Note: if the coefficients of f and ν are p-adic integers, the theorem would still hold provided p is odd, the r_i are in $R(p)$ and a_1r_1 is a unit.

Thus if $f - \nu x_{n+1}^2 = 0$ has a non-trivial solution with $x_{n+1} = 0$, f is a zero form and hence represents ν. This yields

COROLLARY 13. *In any field F, $f = \nu$ has a non-trivial solution if and only if $f - \nu x_{n+1}^2$ is a zero form.*

This shows that for a field the determination of conditions that a form be a zero form is equivalent to finding the numbers of the field represented by forms of one fewer variables.

15. **The solvability of $f = \nu$.** We can now establish criteria for the solvability of $f = \nu$ where ν and the coeffi-

cients of f are in any p-adic field, p being finite. By theorem 11 we may assume without loss of generality that the coefficients of f are rational integers and that ν is a rational integer which we therefore replace by N. These restrictions are assumed for this section, n is taken to be the number of variables in f and d its non-zero determinant.

THEOREM 14a. *If $n = 2$, then f is a zero form in $F(p)$ if and only if $-d$ is a square number in $F(p)$.*

This is true since without altering d or the fact that f is a zero form, we may, by theorem 11, take f into $a_1x_1^2 + a_2x_2^2$. If r_1, r_2 is a non-trivial solution of $f = 0$, neither r can be zero and we have $-d = -a_1a_2 = (a_2r_2/r_1)^2$. Conversely if d is a square we have a non-trivial solution. Corollary 13 yields

COROLLARY 14a. *The equation $ax^2 = N \neq 0$ is solvable if and only if aN is a square.*

THEOREM 14b. *If $n = 3$, f is a zero form in $F(p)$ if and only if $c_p(f) = 1$.*

We may, without loss of generality write $f = a_1x_1^2 + a_2x_2^2 + a_3x_3^2$. To say that $f = 0$ has a non-trivial solution r_1, r_2, r_3 with $r_2 \neq 0$ is equivalent to saying that

$$(-a_1/a_2)(r_1/r_2)^2 + (-a_3/a_2)(r_3/r_2)^2 = 1$$

in other words that $-a_1a_2x_1^2 - a_2a_3x_3^2 = 1$ has a solution. Hence, by the definition of the Hasse symbol, f is a zero form if and only if $(-a_1a_2, -a_2a_3)_p = 1$. The left side of this equation was shown in section 11 to be equal to $c_p(f)$.

Using property 4′ of the Hasse symbol we have, for f a binary form

$$c_p(f - Nx^2) = c_p(f)(-d, -N)_p .$$

Hence, using corollary 13 we have

COROLLARY 14b. *If f is a binary form of determinant d*

and leading coefficient a_{11}, *then* $f = N$ *is solvable for* $N \neq 0$ *if and only if*

$$c_p(f) = (-d, -N)_p ,$$

that is,

$$(-a_{11}, -d)_p = (-N, -d)_p .$$

THEOREM 14c. *If* $n = 4$, f *is a zero form in* $F(p)$ *if and only if the following condition holds:* $c_p(f) = 1$ *when* d *is a square in* $F(p)$.

First we take f into a simpler form f' which is a zero form if and only if f is, and which satisfies the condition of the theorem if and only if f does. By theorem 11 we may take f in the form $a_1x_1^2 + a_2x_2^2 + a_3x_3^2 + a_4x_4^2$ where the coefficients are integers. Replacing a_i by a_ip^{-2} does not alter $c_p(f)$ or its being a zero form; hence we may assume that no coefficient is divisible by p^2. If every a_i is divisible by p, replace f by f/p, if all but the first coefficient is divisible by p, replace x_1 by px_1, multiply a_1 by p^2 and divide the form by p. These operations do not alter the "zeroness" of the form and, since by property 3 of the Hasse symbol, $c_p(pf) = (p, d)_pc_p(f)$ we see that f and f/p both satisfy or both do not satisfy the condition of the theorem. Thus we confine our attention to forms

$$f = a_1x_1^2 + a_2x_2^2 + a_3x_3^2 + a_4x_4^2$$

where a_1a_2 is prime to p, a_3 and a_4 are not divisible by p^2 and all coefficients are integers. For what follows write $f = g + a_4x_4^2 = a_1x_1^2 + h$ and, using property 4′ of the Hasse symbol see that

$$c_p(f) = c_p(g)(d, a_4)_p = c_p(h)(d, a_1)_p .$$

Below we omit the subscripts on the Hasse and Hilbert symbols.

Consider p odd. If $a_1a_2a_3$ is prime to p, the coefficients of g are prime to p and by theorem 14b and property 1 of the Hasse symbol g and hence f is a zero form; furthermore either d is prime to p in which case $c(f) = 1$ or $a_4 = pa_4'$ with a' prime to p in which case d is a non-square. It remains to consider $a_3 = pa_3'$, $a_4 = pa_4'$ where $a_3'a_4'$ is prime to p. Then $(d, a_1) = 1$ and

$$c(f) = c(h) = (a_3, -a_3a_4) = (-a_3'a_4' \mid p).$$

Thus if $c(f) = 1$, then $c(h) = 1$ and h and thus f is a zero form. If $c(f) = -1$ and d is a non-square, $-1 = (-a_3'a_4' \mid p) = (a_1a_2a_3'a_4' \mid p)(-a_1a_2 \mid p) = -(-a_1a_2 \mid p)$ and by theorem 14a, $a_1x_1^2 + a_2x_2^2$ is a zero form and hence f is. Finally if $c(f) = -1$ and d is a square we have shown that

$$-1 = (-a_3'a_4' \mid p) = (-a_1a_2 \mid p).$$

But if f were a zero form then $f = 0$ would have a solution $\rho_1, \rho_2, \rho_3, \rho_4$ with one ρ_i a unit. Hence $f \equiv 0 \pmod{p^2}$ would have a solution r_1, r_2, r_3, r_4 in integers with at least one r prime to p. But $(-a_1a_2 \mid p) = -1$ implies $r_1 = pr_1'$, $r_2 = pr_2'$ and, dividing the congruence by p, we see that $-1 = (-a_3'a_4' \mid p)$ would imply r_3 and r_4 divisible by p contrary to our supposition about r_i. This completes the proof for p odd.

Consider $p = 2$. First if all coefficients of f are odd section 13 shows that $c(f) = 1$ or -1 according as $4 - i = 1$ or 2 or $4 - i = 0, 3$ or 4, where i is the number of coefficients congruent to $1 \pmod 4$. Thus f satisfies the condition of the theorem unless

$$a_1 \equiv a_2 \equiv a_3 \equiv a_4 \pmod 4, \qquad d \equiv 1 \pmod 8.$$

On the other hand $c(g) = c(h) = -1$ if and only if $a_1 \equiv a_2 \equiv a_3 \equiv a_4 \pmod 4$, that is, f is a zero form unless the coefficients are congruent $\pmod 4$. For such a form

$f \equiv 0 \pmod 8$ is solvable with one variable odd and hence f is a zero form by theorem 9b if and only if the sum of the coefficients is divisible by 8, that is, if and only if exactly three are congruent (mod 8), hence if and only if d is not congruent to 1 (mod 8).

Second if $a_4 \equiv 2 \pmod 4$ and $a_1a_2a_3$ is odd, d is a non-square and f satisfies the condition of the theorem. We must then prove f a zero form. This is done by observing that $f(1, 1, 0, 1)$, $f(1, 1, 2, 1)$, $f(1, 1, 0, 0)$, $f(1, 1, 2, 0)$ are all even and no two congruent (mod 8). Hence one is divisible by 8 which, by theorem 9b, shows that f is a zero form.

Finally suppose a_1a_2 odd and $a_3 \equiv a_4 \equiv 2 \pmod 4$. We first establish a duality between the first pair of coefficients and the last pair by noting that the transformation $x_1 = 2x_1'$, $x_2 = 2x_2'$, $x_3 = x_3'$, $x_4 = x_4'$ takes f into $2f'$ with $f' = 2a_1x_1'^2 + 2a_2x_2'^2 + a_3'x_3'^2 + a_4'x_4'^2$. By theorem 12 and property 3 of the Hasse symbol $c(f) = c(2f') = (2, d)c(f')$ and hence consideration of f' is equivalent to consideration of f. If $a_1a_2 \equiv 1 \pmod 4$, the transformation $x_1 = -a_2x_1'/a_1 + x_2'$, $x_2 = x_1' + x_2'$, $x_3 = x_3'$, $x_4 = x_4'$ takes $\frac{1}{2}f$ into $f'' = \frac{1}{2}a_2(a_2/a_1 + 1)x_1'^2 + \frac{1}{2}(a_1 + a_2)x_2'^2 + a_3'x_3'^2 + a_4'x_4'^2$ whose coefficients are odd numbers and, as above, $c(f) = (2, d)c(f'')$ shows that our proof reduces to a previous case. If $a_1a_2 \equiv -1 \pmod 8$ $a_1x_1^2 + a_2x_2^2 \equiv 0 \pmod 8$ has a solution $(1, 1)$ and hence f is a zero form; but $c(f) = c(g)(d, a_4) = (d, a_4)$ and f satisfies the condition of the theorem. By the duality it remains to consider $a_1a_2 \equiv a_3'a_4' \equiv 3 \pmod 8$. Then $f \equiv 0 \pmod 8$ has no solution with one of the variables a unit and $d/4 \equiv 1 \pmod 8$ which implies $c(f) = (2 \mid -a_1a_2) = -1$. Hence f does not satisfy the condition of the theorem and it is not a zero form. This completes our proof.

As in the previous cases we have the accompanying

COROLLARY 14c. *If f is a ternary form in $F(p)$ of determinant d and N a number in $F(p)$, then $f = N$ is solvable if and only if the following condition holds:*

$$c_p(f) = +(-d, -N)_p \text{ whenever } -dN \text{ is a square.}$$

Our discussion is completed by the proof of

THEOREM 14d. *If $n \geqq 5$, f is a zero form in $F(p)$.*

We need merely show this for $n = 5$. Then, by the discussion of the first part of the proof of theorem 14c (omitting the portion involving $c_p(f)$) we see that we may write

$$f = a_1x_1^2 + a_2x_2^2 + a_3x_3^2 + a_4x_4^2 + a_5x_5^2$$

where the coefficients are integers none of which is divisible by p^2 and no more than two of which are divisible by p.

If p is odd, f contains a ternary form g whose determinant is prime to p. Thus g and therefore f is a zero form.

If $p = 2$, by theorem 14c, f will contain a quaternary form which is a zero form except perhaps when every quaternary form has a determinant which is a square. This would require that the product of every pair of coefficients of f be a square and that all five coefficients are odd. Thus, by the discussion for the case $p = 2$ with all coefficients odd in the proof of theorem 14c, we reduce consideration to

$$f \equiv a_1(x_1^2 + x_2^2 + x_3^2 + x_4^2 + x_5^2) \pmod{4}.$$

Then one of (1, 1, 1, 1, 2), (1, 1, 1, 1, 0) is a solution of $f \equiv 0 \pmod 8$ and f is a zero form.

COROLLARY 14d. *If f is an n-ary form with coefficients in $F(p)$ and $n \geqq 4$, then $f = N$ is solvable in $F(p)$ for any number N in $F(p)$.*

We collect the results of these theorems in

THEOREM 14. *If f is an n-ary form with coefficients in $F(p)$, p finite, and non-zero determinant d, it is a zero form if and only if*

a) *For $n = 2$, $-d$ is a square number.*
b) *For $n = 3$, $c_p(f) = 1$.*
c) *For $n = 4$, $c_p(f) = +1$ whenever d is a square.*
d) $n \geqq 5$.

Furthermore for $n > 2$, f is a zero form in $F(p)$ for all odd primes p not dividing d. (For a generalization of this theorem see corollary 18b.)

COROLLARY 14. *If f is an n-ary form with coefficients in $F(p)$, p finite, and non-zero determinant d then, for any p-adic number N, $f = N$ is solvable in $F(p)$ if and only if*

a) *For $n = 1$, dN is a square.*
b) *For $n = 2$, $c_p(f) = (-d, -N)_p$.*
c) *For $n = 3$, $c_p(f) = (-d, -N)_p$ whenever $-dN$ is a square.*
d) $n \geqq 4$.

Furthermore for $n > 1$, $f = N$ is solvable in $F(p)$ for every prime p not dividing $2dN$.

16. **Congruence of forms.** We showed in section 11 that if two forms f_1 and f_2 are congruent in $F(p)$ condition (6) must hold. We can now easily prove the converse of this statement for p a finite prime.

THEOREM 15. *Let f_1 and f_2 be two forms with coefficients in $F(p)$ for p finite, whose determinants are d_1 and d_2 with $d_1 d_2 \neq 0$ and containing n_1 and n_2 variables respectively; then f_1 is congruent to f_2 if*

$$(10) \quad d_1 = \tau^2 d_2, \qquad n_1 = n_2 \quad \text{and} \quad c_p(f_1) = c_p(f_2),$$

where τ is a p-adic number. (This is the converse of theorem 12.)

The theorem is obvious for $n_1 = n_2 = 1$. Then, in order

to prove the theorem by induction we assume it true for $n_1 = n_2 = k \geqq 1$ and seek to prove it for $n_1 = n_2 = k + 1$.

Let N be a non-zero number represented by f_1. Then the conditions of corollary 14 hold for f_1 and N and, in virtue of (10), they then hold for f_2 and N, which shows that f_2 also represents N. Let $r_1, r_2, \cdots, r_{k+1}$ be a solution of $f_1 = N$. By permuting variables we may assume $r_1 \neq 0$ and form a matrix whose first column is $r_1, r_2, \cdots, r_{k+1}$, whose principal diagonal is $r_1, 1, 1, \cdots, 1$ and whose other elements are zero. This will take f_1 into a form whose leading coefficient (that is, the coefficient of x_1^2) is N. Then, as in the proof of theorem 1 we may eliminate all other terms containing x_1. Thus we have transformations of determinants τ_1 and τ_2 taking f_1 and f_2, respectively, into

$$Nx_1^2 + g_1 \quad \text{and} \quad Ny_1^2 + g_2,$$

where g_1 and g_2 are quadratic forms in k variables. But $d_i = \tau_i^{-2} N \mid g_i \mid$ and $c_p(f_i) = c_p(g_i)(d_i, N)_p$ show, in virtue of (10) that $\mid g_1 g_2 \mid$ is a square and $c_p(g_1) = c_p(g_2)$. Thus, by the hypothesis of our induction g_1 and g_2 are congruent. Thus $Nx_1^2 + g_1$ and $Ny_1^2 + g_2$ are congruent and hence f_1 and f_2 are congruent.

17. **Representation of one form by another.** In view of theorem 5 we can solve the problem of determining the forms in m variables represented in $F(p)$, p finite, by forms in n variables, $n > m$, if we can determine to what extent forms can be constructed with given invariants. (The special case $m = 1$ has already been investigated.) This determination is supplied by the following theorem.

THEOREM 16. *Given a positive integer n, a p-adic non-zero number d and a value $+1$ or -1, then there is a form f in $F(p)$, p finite or $p = \infty$, having n variables, determinant d, and the given value for $c_p(f)$ with the following restrictions*

1. *If* $n = 1$, $c_p(f) = (-1, -d)_p$.
2. *If* $n = 2$ *and* $-d$ *is a square, then* $c_p(f) = 1$.

To prove this notice that if $n = 1$, $c_p(f)$ must have the value in restriction 1 above. If $n = 2$, $c_p(f) = (-a_1, -d)_p$ where a_1 is the leading coefficient of f. Then, by property 9 of the Hilbert symbol, if $-d$ is not a square, we may choose a_1 so that the symbol has the value $+1$ or -1 as we please; then the form $a_1x_1^2 + (d/a_1)x_2^2$ will have the required invariants. If $n = 3$ take $f = a_1x_1^2 + a_2x_2^2 + a_3x_3^2$ and $c_p(f) = (-a_1a_2, -d)_p(-a_1a_2, a_1)_p$. If $-d$ is a square, choose $-a_1a_2$ a non-square and, by property 9 of the Hilbert symbol a_1 so that $(-a_1a_2, a_1)_p = c_p(f)$ has the required sign. If $-d$ is a non-square, choose $-a_1a_2$ so that $(-a_1a_2, -d)_p = -1$. This implies that $-a_1a_2$ is a non-square and a_1 may be chosen so that $c_p(f)$ has the required sign. The cases $n > 3$ are taken care of by noting that $c_p(x_1^2 + g) = c_p(g)$ and the determinants of $x_1^2 + g$ and g are equal. This discussion applies equally to p finite or p infinite.

Now, using theorem 5 we prove

THEOREM 17. *If f and g are forms in $F(p)$ with n_f and n_g variables and non-zero determinants d_f and d_g respectively, with $n_f > n_g$; then f represents g in $F(p)$ with the following restrictions (p being finite):*

1. *If* $n_f - n_g = 1$, *then* $c_p(f)c_p(g) = (-d_g, d_f)_p$.
2. *If* $n_f - n_g = 2$ *and* $-d_gd_f$ *is a square, then* $c_p(f)c_p(g) = (-1, -d_g)_p$.

If $n_f > n_g$ we must show that there is a form h such that $f = g + h$ under the given restrictions. Using property 4 of the Hasse symbol we see that

$$c_p(h) = c_p(f)c_p(g)(-1, -1)_p(d_g, d_gd_f)_p. \tag{11}$$

If $n_f - n_g = 1$, theorem 16 requires that $c_p(h) =$

$(-1, -d_h)_p = (-1, -d_g d_f)_p$ and substitution in (11) gives us condition 1.

If $n_f - n_g = 2$ and $-d_h$ is a square, that is, if $-d_f d_g$ is a square, then from (11), $c_p(h) = 1$ if and only if

$$c_p(f)c_p(g) = (-1, -1)_p(d_g, -1)_p = (-1, -d_g)_p .$$

If $n_f - n_g > 2$, theorem 16 shows that h may always be determined.

This completes the proof of the theorem. The corresponding result for the field of real numbers was proved in chapter I. An immediate consequence of the theorem is

COROLLARY 17. *If p is an odd prime not dividing $|f|$ nor $|g|$ and if $n_f > n_g$, then f represents g in $F(p)$.*

Notice that theorem 7 regarding automorphs and representations holds for $F(p)$ and hence gives us a relationship between any two representations.

A generalization of the notion of a zero form is the following: An n-ary form is called an *m-zero* form in a field F if its matrix A is non-singular and if there is an n by m matrix T in F of rank m such that $T^T A T$ is the zero m by m matrix. In this connection we prove

THEOREM 18. *A non-singular n by n matrix A in a field F in which $2 \neq 0$ is the matrix of an m-zero form if and only if A is congruent in F to a matrix $A_1 \dotplus A_2$ where A_1 is the direct sum of m matrices of binary zero forms $2xy$. Furthermore $m \leqq n/2$.*

First suppose there is a matrix T of the required characteristics taking A into the zero matrix B. By lemma 3 we can find a matrix T_0 such that $(T\ T_0)$ is non-singular. Then $(T\ T_0)$ will take A into a non-singular matrix C whose leading m by m minor is B. Now if $n < 2m$, the first m rows of C are linearly dependent contrary to the assumed non-singularity of A. In C there is one element

$c_{ij} \neq 0$ with $i \leqq m, j > m$. Without loss of generality call it $c_{1,m+1}$. Make the cyclic permutation of columns and rows $(m + 1, 2, 3, \cdots, m)$ and have as the leading 2 by 2 minor

$$\begin{bmatrix} 0 & c_{1,m+1} \\ c_{1,m+1} & c_{m+1,m+1} \end{bmatrix}$$

which is non-singular and the matrix of a binary zero form. Add appropriate multiples of the first column to the other columns and of the first row to other rows to make the second row and column of the leading $m + 1$ by $m + 1$ minor $(c_{1,m+1}, c_{m+1,m+1}, 0, 0, \cdots, 0)$. This will not alter the rest of this principal minor, that is, the elements in the $3, 4, \cdots, m + 1$ rows and columns are all zero. Then we may add appropriate multiples of the first two rows and columns to the $m + 2, \cdots, n$ rows and columns to make the first two rows and columns of C all zero except for the leading 2 by 2 minor. Then the matrix C is the direct sum of the 2 by 2 minor above and a matrix C' whose leading $m - 1$ by $m - 1$ minor is a zero matrix. So we may proceed to prove our theorem.

On the other hand suppose $A = A_1 \dotplus A_2$ where A_1 is the direct sum of matrices of binary zero forms. Each matrix of a binary zero form may be assumed to be of the form

$$\begin{bmatrix} 0 & c_1 \\ c_1 & c_2 \end{bmatrix}.$$

Permute the columns and rows of A_1 so that the third column becomes the second, the fifth becomes the third, $\cdots$, the $(2m - 1)$th becomes the m-th and similarly for the rows. The leading m by m minor of the result is the zero matrix.

It remains to show that every binary zero form is congruent to $2xy$. Since such a form represents zero non-trivially it is congruent to $2cxy + c_2y^2$ for $c_1 \neq 0$. The transformation $2x' = 2c_1x + c_2y$, $y' = y$ takes this into $2x'y'$.

If F is the field of real numbers, the index of the direct sum of binary forms is m. Then theorem 6 gives us

COROLLARY 18a. *A form f of index i and rank n in the field of reals is an m-zero form if and only if*

$$n - m \geqq i \geqq m.$$

Now suppose F is the field of p-adic numbers. If $n = 2m$, theorems 12 and 15 tell us that f is congruent to g, a sum of binary zero forms if and only if $c_p(f) = c_p(g)$ and $|f| \sim |g|$. But since g may be taken to be the sum of terms $2x_{2i-1}x_{2i}$ we see that $c_p(g) = 1$ for p odd or, using section 13, $c_p(g) = (-1)^{(m-2)(m-1)/2}$ for $p = 2$. For $n > 2m$, use theorem 17 to prove

COROLLARY 18b. *An n-ary form with non-zero determinant d in $F(p)$ is an m-zero form if and only if $n \geqq 2m + 3$ or one of the following holds, p being finite,*

1. *If $n = 2m$, $|f| \sim (-1)^m$ and $c_p(f) = w$ where $w = 1$ or $(-1)^{(m-2)(m-1)/2}$ according as p is odd or $p = 2$.*
2. *If $n = 2m + 1$, $c_p(f) = w(\{-1\}^{m+1}, d)_p$ where w has the value just above.*
3. *If $n = 2m + 2$ and, whenever $d(-1)^{m+1}$ is a square, then $c_p(f) = 1$ or $(-1)^{(m^2-m)/2}$ according as p is odd or $p = 2$.*

18. **Universal forms.** Though we have now solved the principal problems for p-adic fields there is one by-product of section 15 which we shall consider. We call a form *universal* in a field if it represents all the non-zero numbers of the field. In the field of reals, a form is universal if and only if it is indefinite. In what follows then we restrict p to

be finite except where otherwise noted. (Notice that every form represents 0 but some represent it only trivially.)

Theorem 13 shows that for $F(p)$ any zero form is universal. The converse of this statement follows with one exception from corollary 14 as may be seen as follows. If $n = 1$ there are no zero or universal forms. If $n = 2$, $(-d, -N)_p$ has, by property 9 of the Hilbert symbol, the same value for all N if and only if $-d$ is a square, that is, using theorem 14, if and only if f is a zero form. If $n = 3$ take $N = -d$, see that $-dN$ is a square and hence by corollary 14, $f = -N$ solvable implies $c_p(f) = (d, -d)_p = 1$ and thus by theorem 14, f is a zero form. If $n \geqq 4$ all forms are universal in $F(p)$ for p finite and all forms are zero forms except for $n = 4$, $c_p(f) = -1$ and d a square.

Thus we have

THEOREM 19. *The totality of zero forms coincides with the totality of universal forms in $F(p)$ for p finite or $p = \infty$ except that under the following conditions a form f of determinant d is universal but not a zero form:*

$$p \text{ finite}, \qquad n = 4, \qquad c_p(f) = -1, \qquad d \text{ a square}.$$

An example of a universal form not a zero form is

$$x_1^2 + 2x_2^2 + 5x_3^2 + 10x_4^2$$

in $F(5)$. (See theorem 84).

We have a generalization of the idea of a universal form which is analogous to our generalization of a zero form. If A is the matrix of an n-ary form f and if, for every m by m symmetric matrix B of rank m there is a matrix T such that $T^TAT = B$ we call f an *m-universal form*. We prove the following

THEOREM 20. *If F is the field of real numbers, then a form f in F is an m-zero form if and only if it is m-universal. The* *"only if" part of the theorem holds for F the field of p-adic numbers.*

First suppose f is an m-zero form. Then, by theorem 18, it is congruent to $f_1 + f_2$ where f_1 has matrix C and is the direct sum of m binary zero forms. Thus we wish to show that $C = C_1 \dotplus C_2 \dotplus \cdots \dotplus C_m$ represents an arbitrary symmetric matrix $B = (b_{ij})$ of m rows. Now each C_i has non-zero determinant and since it is the matrix of a universal binary form may be taken to have its leading element b_{ii}. From Theorem 1 there is a non-singular matrix T such that $T^T BT = B_1$ is diagonal. Hence f represents B if and only if it represents B_1 and we may without loss of generality, take B to be diagonal. Then, crossing out the even numbered rows and columns of C leaves the matrix B. This shows that C represents B.

Second, suppose F is the field of real numbers and f is an m-universal form. Then theorem 6 tells us that $i \geqq j$ and $n - i \geqq m - j$ for all $j \leqq m$. Hence $i \geqq m$ and $n - i \geqq m$ which is by corollary 18a the condition that f be an m-zero form.

It is not hard to find the conditions that f be m-universal in $F(p)$. Theorem 17 shows us that f is m-universal if $n > m + 2$. First if $n = m + 2$, the same theorem shows us that the condition that f be universal is that

$$(12) \qquad c_p(f) = (-1, -d_g)_p c_p(g) = (-1, d_f)_p c_p(g),$$

for every g for which $-d_g d_f$ is a square. Assume that d_g is chosen so that $-d_g d_f$ is a square. If $m = 1$, theorem 16 shows that the choice of g implies $(-1, -d_g)_p c_p(g) = 1$. If $m = 2$, g may be chosen to deny (12) unless $-d_g$ is a square in which case d_f is a square and $c_p(f) = c_p(g) = 1$. If $m > 2$ the equality (12) may be denied.

Second if $n = m + 1$ theorem 14 suffices for $m = 1$ and for $m \geqq 2$, after having chosen $-d_g$ a non-square, we can choose $c_p(g)$ so that $c_p(f) \neq (-d_g, d_f)_p c_p(g)$ and see that f is not m-universal.

Third if $n = m$ we have the choice of either $c_p(g)$ or d_g for all values of m and hence no form is m-universal. These results are included in the theorem below.

THEOREM 21. *An n-ary form f in $F(p)$ is m-universal if $n > m + 2$ or if one of the following holds*:

1. $n = m + 2$ *and either* $m = 1 = c_p(f)$ *or* $m = 2$ *with* $d_f \sim 1 = c_p(f)$.
2. $n = m + 1$ *and* $m = 1$ *with* $-d_f \sim 1$.

Superficially it may seem strange that theorem 20 does not give a necessary and sufficient condition for $F(p)$ as well as for the field of reals. If we had defined an m-universal form to be one for which $T^T A T = B$ has, for every B, a solution T in $R(p)$ where T has an m-rowed minor which has unit determinant then theorem 9b would show that A would be an m-zero form since we could make the elements of B divisible by an arbitrarily high power of p.

Chapter III

FORMS WITH RATIONAL COEFFICIENTS

19. **Congruence.** When two forms with rational coefficients may be taken into each other by linear transformations with rational elements, the forms are frequently called rationally equivalent. But, to be consistent with our terminology, we shall call two such forms *rationally congruent* or *congruent* in the field of rational numbers and reserve the term "equivalent" for transformations with coefficients in a ring. We shall see that there is an intimate connection between the fundamental results of this chapter and those of the previous chapter.

Since the rational numbers form a field we have shown in theorem 1 that every form is rationally congruent to a diagonal form. As in the last chapter, we can specialize this still further; for $(r/s)x^2$, where r and s are integers, becomes rsy^2 if x is replaced by sy and any square factor of a coefficient may be absorbed into the variable. Hence we have

THEOREM 22. *Every form with rational coefficients is rationally congruent to a diagonal form whose coefficients are square-free integers (that is, integers with no square factors except* 1).

20. **Equivalence and reduced forms.** So far in this book we have considered transformations whose elements are in the same field as the coefficients of the form. Here, as is perhaps suggested by theorem 22, there is some profit in considering a more restricted type of transformation, namely one with integer elements. If such a transformation is to have an inverse which also has integer

elements its determinant must be ± 1 since $TT^I = I$, the identity, implies that the product of the determinants of T and T^I is 1, whereas $|T|$ and $|T^I|$ are both integers. Thus we define a *unimodular transformation* to be one which has integral elements and determinant ± 1. In classical literature this term is frequently reserved for transformations of determinant $+1$. However we shall use the term *properly unimodular* for a transformation with integer elements and determinant $+1$. In the case of transformations with elements in a ring but not a field, it is customary to replace the term congruence by equivalence and hence we refer to two forms f and g as being *equivalent* or in the same *class* if one may be taken into the other by a unimodular transformation; we denote this by $f \cong g$. On occasion we find it convenient to write $a \cong b$ for two numbers of $F(p)$ which means that a/b is the square of a unit. If the determinant of the transformation taking f into g is $+1$ we say that the two forms are *properly equivalent* and in the same proper class. If the determinant of the transformation is -1 the term "improper" is used in place of "proper". Notice that two equivalent forms must have the same number of variables and the same rank but need not have non-zero determinant. However, for the most part, we consider only forms with non-zero determinant.

The canonical form for an equivalence transformation is called a *reduced form*. We give in the following theorem a reduced form essentially due to Hermite.

THEOREM 23. *If g is a form in n variables with rational coefficients and non-zero determinant d and if g is not a zero form, then g is equivalent to a form $f = \Sigma a_{ij}x_ix_j$ having the following properties:*

1. $0 < |a_{11}| \leqq (4/3)^{(n-1)/2} \sqrt[n]{|d|}$,
2. $|a_{11}| \geqq 2|a_{1j}|, j > 1$,

3. $a_{11}f - (a_{11}x_1 + a_{12}x_2 + \cdots + a_{1n}x_n)^2 = f_1$

where f_1 *is a form in* $n - 1$ *variables satisfying, with* n *replaced by* $n - 1$, *the same conditions imposed on* f. *The determinant of* f_1 *is* da_{11}^{n-2}.

The theorem trivially holds for $n = 1$. Hence to prove it by induction we assume it for $n = k - 1$ and show it for $n = k$.

Let $a_{11} \neq 0$ be some number represented by g for integral values of the variables and $c_1, c_2, \cdots, c_k$ be a solution of $g = a_{11}$. If the g.c.d. of the c's is w, then $(c_1/w, c_2/w, \cdots, c_k/w)$ is a solution of $g = a_{11}/w^2$. Thus we may replace a_{11} by a_{11}/w^2 and assume that the c's have 1 as their greatest common divisor. Then form a matrix T of determinant 1 whose elements are integers (see lemma 6 proved below) and whose first column is $c_1, c_2, \cdots, c_k$. This takes g into a form f whose leading coefficient is a_{11}.

Consider the transformation

$$\begin{aligned} x_1 &= y_1 + c_{12}y_2 + \cdots + c_{1k}y_k \\ x_2 &= \qquad c_{22}y_2 + \cdots + c_{2k}y_k \\ &\cdots\cdots\cdots\cdots\cdots\cdots \\ x_k &= \qquad c_{k2}y_2 + \cdots + c_{kk}y_k . \end{aligned}$$

This will take f into $F = \Sigma A_{ij}y_iy_j$ with $A_{11} = a_{11}$. Write

$$a_{11}F - (a_{11}y_1 + A_{12}y_2 + \cdots + A_{1k}y_k)^2 = F_1$$

where F_1 is free of y_1. Now, by the hypothesis of the induction, we can choose c_{ij}, $i, j > 1$, so that $| c_{ij} | = \pm 1$ and F_1 satisfies the conditions of the theorem since F_1 is a form in $k - 1$ variables. However, F_1 contains only $y_2, y_3, \cdots, y_k$ whose values in terms of $x_2, x_3, \cdots, x_k$ depend only on c_{ij} for $i, j > 1$. Hence F_1 is independent of

$c_{12}, \cdots, c_{1k}$, that is, we are free to determine these c's as we please without altering F_1. Now the only terms containing y_1 come from $a_{11}x_1^2 + 2a_{12}x_1x_2 + \cdots + 2a_{1k}x_1x_k$ which becomes

$$a_{11}y_1^2 + 2(a_{11}c_{12} + \Sigma\, a_{1j}c_{j2})y_1y_2 \\ + \cdots + 2(a_{11}c_{1k} + \Sigma\, a_{1j}\, c_{jk})y_1y_k +$$

terms free of y_1. Thus the c_{1i} may be determined so that conditions 2 of the theorem hold for the coefficients of y_1y_i, $i > 1$. Hence we may consider the coefficients of $f_1 = F_1$ to satisfy conditions 2 of the theorem.

Property 1 remains to be established. Again, it is trivial for $n = 1$ and we assume it for $n = k - 1$, that is, that if b is the leading coefficient of f_1, then

$$|\, b \,| = |\, a_{11}a_{22} - a_{12}^2 \,| \leqq (4/3)^{(k-2)/2} \sqrt[k-1]{|\, d_1|}$$

where d_1 is the determinant of f_1. The determinant of $a_{11}f$ is $a_{11}^k d$ and hence $da_{11}^k = a_{11}^2 d_1$. If a_{22} were zero, $(0, 1, 0, \cdots, 0)$ would be a non-trivial solution of $f = 0$ and it would be a zero form. If $|\, a_{11} \,|$ is greater than $|\, a_{22} \,|$ we interchange x_1 and x_2 in f and start over again to satisfy condition 2 for the new coefficient. This process cannot go on forever since all numbers represented by the form with integer values of the variables are integers divided by the l.c.m. of the denominators of the coefficients of the form, that is, given any number M, there is only a finite number of positive numbers less than M represented by the form. Thus, after a finite number of back-trackings we may take $|\, a_{11} \,| \leqq |\, a_{22} \,|$ and, by condition 2, $|\, a_{11} \,| \geqq |\, 2a_{12} \,|$ which imply $a_{11}^2 \leqq a_{11}^2/4 + b$ or

$$3a_{11}^2/4 \leqq |\, b \,| \leqq (4/3)^{(k-2)/2}(|\, a_{11}^{k-2} d|)^{1/(k-1)}.$$

This implies the inequality in 1 and completes the proof.

One might be tempted to try this theorem for forms with

real coefficients but, though condition 2 could be made to hold, the proof would fail at an essential point, namely the finiteness of the number of back-trackings referred to. Consider for instance the form $x^2 - \sqrt{2}y^2$. The theory of continued fractions can be used to show that this can be made arbitrarily small for integer values of x and y. The theorem can be modified to hold for real coefficients but it is not here worth our while.

As an application of this theorem, suppose $d = \pm 1$, $n = 5$ and the coefficients of the form are integers. Since, for integer values of the variables, f represents only integers, condition 1 shows that we may take $a_{11} = \pm 1$ and from condition 2, $a_{1i} = 0$, $i = 2, 3, 4, 5$. Thus f_1 has determinant ± 1 and the same reasoning may be applied to it. Thus f reduces to diagonal form with coefficients 1 and -1. If two coefficients were of different sign the form would be a zero form which is excluded by the theorem. Thus we might just as well restrict ourselves to positive forms and have

COROLLARY 23. *If f is a positive form in fewer than six variables and having integral coefficients and determinant* 1, *it is equivalent to*

$$x_1^2 + x_2^2 + \cdots + x_n^2 .$$

As a matter of fact, sharper inequalities than 1 imply the same result for $n = 6$ and $n = 7$. It is not true for $n \geqq 8$.

We shall prove later on in this chapter that every indefinite form with more than four variables and rational coefficients is a zero form. Thus our theorem applies only to definite forms and to some indefinite forms of less than five variables.

Before finding a canonical form for zero forms we need three lemmas.

LEMMA 4. *Let T be an n by m matrix with "integer" elements, where in this lemma and its proof "integer" means rational integer or integers in a p-adic field. There are unimodular matrices P and Q (matrices with "integer" elements and unit determinant) such that*

1. *PTQ is a matrix (s_{ij}) where $s_{ij} = 0$ if $i \neq j$ and s_{ii} divides s_{jj}, $j \geqq i$.*
2. *P and Q are products of matrices of type E_1 obtained by interchanging two rows or columns of the identity matrix and of type E_2 obtained by adding an "integral" multiple of one row (or column) to another row (or column) of the identity matrix.*

To prove this lemma we show that T may be reduced to a diagonal form by a series of permutations of rows or columns and adding integral multiples of a row (or column) to another row (or column). Find the "smallest" element of T where "smallest" means of least positive value for rational integers and containing the least power of p for p-adic integers; by a permutation of rows and columns assume it to be t_{11}. Then, for each t_{1j}, $j > 1$, we can find an integral element r_j such that $t'_{1j} = t_{11}r_j + t_{1j}$ is zero or a smaller integer than t_{11} and by adding r_j times the first column to the j-th we have all but the first element of the first row zero or some element in the first row smaller than t_{11}. In the latter case permute columns to put the smallest element in the t_{11} position and repeat the process. Thus eventually we will have $t_{11}, 0, \cdots, 0$ as the first row. Then proceed similarly with the elements of the first column to make the first row and column identical. If now we carry through the same process for the matrix obtained by deleting the first row and column we get a matrix (s_{ij}) with $s_{ij} = 0$ for $i \neq j$ and $s_{ii} \neq 0$ if i is not greater than the rank of the matrix. If now s_{11} does not divide s_{22} add the second row to the first and then proceed as above to replace s_{11} by a g.c.d. of s_{11} and s_{22}, and so the proof may be completed.

If T is a unimodular matrix the resulting diagonal matrix must have all its elements units. Hence we have the

COROLLARY 24. *If T is a unimodular matrix it is expressible as a product of matrices of types E_1 and E_2 with a diagonal matrix whose elements are units. In other words, any unimodular matrix may be obtained from the identity matrix by the multiplication of its rows by units followed by multiplication by matrices of types E_1 and E_2.*

LEMMA 5. *Let T be an n by m matrix with "integer" elements in the sense of lemma 4 and let $\lambda_r(T)$ be the* g.c.d. *of its r-rowed minor determinants. Multiplying T by a unimodular matrix P leaves $\lambda_r(T)$ invariant for all r.*

From the corollary above we need only show that $\lambda_r(T)$ is left invariant by multiplication by matrices of the three types: E_1, E_2, and multiplication of rows by units. It is obvious that neither the first or the third type alters $\lambda_r(T)$. It remains to examine transformations of type E_2. If A_r is any r-rowed minor determinant of T, adding an integral multiple τ of one row to another or one column to another either leaves A_r unaltered or replaces it by $A_r + \tau B_r$ where B_r is another r-rowed minor determinant. Hence any number dividing all r-rowed minors before divides all r-rowed minors after the transformation. That is, $\lambda_r(T)$ divides $\lambda_r(T')$ where $E_2T = T'$. But $T = E_2^I T'$ shows that $\lambda_r(T')$ divides $\lambda_r(T)$ and hence $\lambda_r(T') = \lambda_r(T)$.

LEMMA 6. *Let T by an n by m matrix with $n > m$ and having integral elements in the sense of lemma 4 such that g is the* g.c.d. *of its m by m minors. There is an n by $n - m$ matrix T_0 with integral elements such that the n by n matrix $(T\ T_0)$ has determinant g.*

To prove this notice that the process used to prove

lemma 4 shows the existence of a unimodular matrix P such that $PT = U = (u_{ij})$ with $u_{ij} = 0$ for $i > j$. Since, by lemma 5, this process does not alter the g.c.d. of the m-rowed minor determinants of T, $g = u_{11}u_{22} \cdots u_{mm}$. Let J be an n by $n - m$ matrix whose last $n - m$ rows constitute the identity matrix. Then the n by n matrix $(U\ J)$ has determinant g. Hence $P^I(U\ J) = (P^I U\ P^I J) = (T\ P^I J)$ has determinant g.

In order to make one direct application of this lemma suppose $(r_1, r_2, \cdots, r_n)$ is a solution in integers of $f = N$ and the g.c.d. of the r_i is 1. Then by the lemma form a unimodular matrix S whose first column is $(r_1, r_2, \cdots, r_n)$. This will take f into a form whose leading coefficient is N. Hence we have the

Corollary. *If $f = N$ has a solution $(r_1, r_2, \cdots, r_n)$ in integers whose g.c.d. is 1, then f is equivalent to a form whose leading coefficient is N.*

THEOREM 23a. *An n-ary zero form f with rational coefficients is equivalent to a form g whose matrix (a_{ij}) is of the form*

$$\begin{bmatrix} N & B \\ B^T & C \end{bmatrix} \tag{13}$$

where N is an m by m zero matrix (that is, all its elements are zero), B is an m by $n - m$ matrix (b_{ij}), with $b_{ij} = 0$ for $i \neq j$ and b_{jj} is an integral multiple of b_{ii} for $j \geqq i$, while C is an $n - m$ by $n - m$ matrix (c_{ij}), in which, for $i \leqq m$, $2 \mid c_{ki} \mid \leqq \mid b_{ii} \mid$ when $k \neq i$ and $\mid c_{ii} \mid \leqq \mid b_{ii} \mid$. The inequality $n \geqq 2m$ holds.

Since f represents zero non-trivially, there is a column matrix whose elements are integers of g.c.d. 1 taking f into 0. Let T be an integral n by m matrix of maximum number of columns, the g.c.d. of whose m-rowed minors is 1 and taking the matrix of f into a zero matrix N. From lemma 6 we can find a matrix T_0 such that $(T\ T_0)$ is unimodular. This will take the matrix f into a matrix (13) with $B = (b_{ij})$ and $C = (c_{ij})$. Let s be the l.c.m. of the denominators of the elements of B and C and write $sB = \bar{B}$, $sC = \bar{C}$. From lemma 4 we can find unimodular matrices P and Q such that $P\bar{B}Q = \bar{B}_0$ where the elements of $\bar{B}_0$ satisfy the conditions of our theorem imposed on B. Then

$$\begin{bmatrix} P & 0 \\ 0 & Q^T \end{bmatrix} \begin{bmatrix} N & \bar{B} \\ \bar{B}^T & \bar{C} \end{bmatrix} \begin{bmatrix} P^T & 0 \\ 0 & Q \end{bmatrix} = \begin{bmatrix} N & \bar{B}_0 \\ \bar{B}_0^T & Q^T\bar{C}Q \end{bmatrix}.$$

By adding appropriate multiples of the first m rows and columns to the remaining rows and columns we can make the elements of Q^TCQ satisfy the requirements on C. Dividing by s all elements of the resulting matrix yields a matrix satisfying the conditions imposed. Since $|f| \neq 0$, the first m rows of (13) must be linearly independent and hence $n - m \geqq m$.

Notice that if f is an integral form, the square of the product $b_{11}b_{22} \cdots b_{mm}$ divides $|f|$ and if $|f|$ is square free, all non-diagonal elements of the leading m by m minor of C may be taken to be zero and all diagonal elements of absolute value 1.

There are three theorems which we include here because of their later usefulness.

THEOREM 24. *If f is an n-ary form with integral coefficients which is a zero form in $F(p)$, there is a transformation of determinant p^{n-1} and having integral elements which takes f into pg where g is a form having integral coefficients.*

To prove this let $r_1, r_2, \cdots, r_n$ be a non-trivial solution in integers of $f \equiv 0 \pmod{p}$. We may, without loss of generality, take $r_n \not\equiv 0 \pmod{p}$ and, dividing through by r_n, have a solution $s_1, \cdots, s_{n-1}, 1$ of $f \equiv 0 \pmod{p}$. Then the transformation whose last column is

$s_1, s_2, \cdots, s_{n-1}, 1$, whose principal diagonal consists of p's except for its last element and whose other elements are zero, does what is required.

THEOREM 25. *If f is a form with integral coefficients, p an odd prime and k an arbitrary positive integer, then there is a form g equivalent to f such that*

$$g \equiv a_1x_1^2 + a_2x_2^2 + \cdots + a_nx_n^2 \pmod{p^k},$$

where we say that one form is congruent to another mod p^k *if corresponding coefficients are congruent.*

Let p^t be the highest power of p dividing all the elements of f. If a_{ii}/p^t is prime to p for some i, we may permute variables so that a_{11}/p^t is prime to p. If $a_{ii} \equiv 0 \pmod{p^{t+1}}$ for all i, let a_{ij}/p^t be prime to p and see that the transformation $x_k = y_k$, $k \neq i$, $x_i = y_i + y_j$ yields a form in which the coefficient of y_j^2 is $b = a_{ii} + 2a_{ij} + a_{jj}$ and b/p^t is prime to p. Hence we may take a_{11}/p^t prime to p.

Then, since a_{1i} are all divisible by p^t we see that the congruences $(a_{11})x + a_{1i} \equiv 0 \pmod{p^k}$ have integral solutions $r_2, r_3, \cdots, r_n$. The transformation whose first row is $1, r_2, \cdots, r_n$, whose diagonal elements are 1, and whose other elements are zero takes f into a form the first row and column of whose matrix is $\equiv (a_{11}, 0, \cdots, 0) \pmod{p^k}$. Then we proceed in like manner with that portion of the form free of x_1. The theorem for $p = 2$ corresponding to theorem 25 is theorem 41.

THEOREM 26. *If f is an n-ary form with integral coefficients and of determinant d, while m is a number such that $c_p(f) = (\pm d, m)_p$ for a preassigned value of the ambiguous sign and for all primes p dividing $2d$ as well as for $p = \infty$, then, for any modulus R whose prime factors are all factors of $2d$, there is an $M \equiv m$* (mod R) *such that $c_p(f) = (\pm d, M)_p$ for all primes p including $p = \infty$.*

Let g be the g.c.d. of m and R. Then by Dirichlet's theorem on the primes in an arithmetic progression we know that there is a prime number P not dividing $2d$ such that $M = gP \equiv m \pmod{R}$ where g is given the sign of m. Then for R sufficiently large we have $(\pm d, M)_p = (\pm d, m)_p$ for all primes p dividing $2d$. If p is a prime not dividing $2dP$ we have $c_p(f) = (\pm d, M)_p = 1$. Thus $c_p(f) = (\pm d, M)_p$ for all primes p except perhaps $p = P$. But, since 1 is the product over all primes p of the symbols on the opposite sides of the last equation the equality holds also for $p = P$ and our theorem is proved.

If f is a binary form, the solvability of $f = m$ in $F(p)$ is equivalent to the condition $c_p(f) = (-d, -m)_p$. Hence we have

COROLLARY 26. *If f is a binary form with integral coefficients and of determinant d and m is a number such that $f = m$ is solvable in $F(p)$ for all primes p dividing $2d$ and $p = \infty$, then for any modulus R whose prime factors are divisors of $2d$, there is an $M \equiv m \pmod{R}$ such that $f = M$ is solvable in $F(p)$ for all primes p. (This is a special case of theorem 40, part 3a.)*

21. **The fundamental theorem on zero forms.**

A zero form in the field of rational numbers is a zero form in $F(p)$ for every prime p including $p = \infty$. The truth of the converse of this statement, given in theorem 27, below is a remarkable result first proved by Hasse. This statement and others similar to it underly much of the remaining theory of this book.

We shall see that for quaternary forms of determinant whose absolute value is a square and for ternary forms, the proof of theorem 27 is independent of theorem 26 and hence of Dirichlet's theorem on primes in an arithmetic progression. As a matter of fact, Pall has shown that results from the composition of binary forms may be used to avoid Dirichlet's theorem altogether. We now prove

THEOREM 27. *If f is a form of non-zero determinant and with rational coefficients, then it is a zero form if and only if it is a p-adic zero form for all primes p including $p = \infty$. (See remarks 1 and 2 after the first corollary.)*

In accordance with the above remarks we need prove only the "if" part of the theorem. Since f is a zero form or a p-adic zero form if and only if any form rationally congruent to it has the same property we may assume from theorem 22 that f is a diagonal form with square-free integral coefficients.

First suppose $n = 2$. Then if f is to be a p-adic zero form for all p, $-d$ must be a square for every prime factor of d. Hence $d = \pm s^2$ where s is an integer. Since $-d$ is a square in $F(2)$ (or in $F(\infty)$) the ambiguous sign cannot be positive. Thus $f = a_1x_1^2 + a_2x_2^2$ where $-a_1a_2 = s^2$ and $f = 0$ has the non-trivial solution $x_1 = s$, $x_2 = a_1$. Notice that we did not need to use the fact that $-d$ is a square in the field of reals nor in $F(2)$ but that we needed one of these facts. This is natural since $\Pi c_p(f) = 1$, the product being over all p, and for $n = 2$, $c_p(f)$ determines the sign of d.

Second suppose $n = 3$ and $f = a_1x_1^2 + a_2x_2^2 + a_3x_3^2$, where the coefficients are square-free integers. We may assume the g.c.d. of the coefficients to be 1 since otherwise we could divide by a common factor and have a form which is a zero form if and only if f is. If p divides a_2 and a_3, $f = 0$ if and only if $x_1 = px_1'$ where x_1' is an integer and in that case we may consider instead of f, the form $a_1px_1'^2 + (a_2/p)x_2^2 + (a_3/p)x_3^2$. Thus we assume that no two coefficients have a factor in common.

Suppose p is an odd prime dividing a_3 but not a_1a_2. Then f a p-adic zero form implies that $c_p(f) = 1$, that is

$(-a_1a_2 \mid p) = 1$. Thus $a_1x_1^2 + a_2x_2^2 \equiv 0 \pmod{p}$ has the solution $r, 1$ and the transformation

$$\begin{bmatrix} p & r & 0 \\ 0 & 1 & 0 \\ 0 & 0 & 1 \end{bmatrix}$$

takes f into pg where $g = a_1py_1^2 + 2a_1ry_1y_2 + by_2^2 + (a_3/p)y_3^2$, $b = (a_1r^2 + a_2)/p$ and the determinant of g is prime to p. Now if $g = 0$ has a non-trivial solution in integers so has f and $|g| = |f|/p$. Suppose q is an odd prime dividing the determinant of g. We see that $q \neq p$ and theorem 25 shows that g is congruent to a form

$$g' \equiv b_1x_1^2 + b_2x_2^2 + b_3x_3^2 \pmod{p^2}$$

where the coefficients are integers and we proceed as above to eliminate the factors of $|g'|$ involving q. So we continue until we have a form h whose determinant is a power of 2 contained in d, whose coefficients are integers and which is a zero form if and only if f is. Now by the discussion at the beginning of this case, the determinant of f is not divisible by 4. Hence the determinant of h is ± 1 or ± 2. Thus either h is a zero form or theorem 23 applies to show that we may take the leading coefficient of h to be $<(4/3)\sqrt[3]{2}$, in absolute value, that is $+1$. Again using theorem 23 we can show that $\pm h = x_1^2 + cx_2^2 + ex_3^2$ where $c = 1$ or -1 and $e = 1, -1, 2$ or -2. Unless c and e are both positive, h is clearly a zero form. If they are both positive, h is not a zero form in the field of reals. Again, examining h in only one of the fields $F(2)$ and $F(\infty)$ is necessary as may be seen directly or by use of property 2 of the Hasse symbol.

Third, suppose $n = 4$. By the argument at the beginning of the last case we see that f may be considered to be in diagonal form with square-free integral coefficients and the g.c.d. of every three equal to 1. Thus d is divisible by no cube except ± 1.

Suppose the last two coefficients of f are even; then the transformation

$$\begin{pmatrix} 2 & 1 & 0 & 0 \\ 0 & 1 & 0 & 0 \\ 0 & 0 & 1 & 0 \\ 0 & 0 & 0 & 1 \end{pmatrix}$$

takes f into a form $2g$ where g has integral coefficients and odd determinant but is not in diagonal form. Thus in what follows we assume the determinant of f not divisible by 4 and though the form is not necessarily in diagonal form it may be assumed diagonal mod p^2 for any odd prime considered.

If p is an odd prime whose square divides d either its square divides one of the coefficients (mod p^2) in which case we may absorb it into the accompanying variable and have a form of determinant prime to p, or p but not p^2 divides just two coefficients. In the latter case assume them the last two and see that $c_p(f) = (-a_3'a_4' \mid p)$ where $a_3 = pa_3'$ and $a_4 = pa_4'$. Thus one of two things must happen. First $c_p(f) = 1$ which implies that $a_3'x_3^2 + a_4'x_4^2$ is a zero form in $F(p)$ and by theorem 24 may be taken into a form pf' where f' is a form $b_1y_3^2 + 2b_2y_3y_4 + by_4^2$ whose coefficients are integers and whose determinant is $a_3'a_4'$. Then $py_3 = z_3$, $py_4 = z_4$ takes p^2f' into a form whose determinant is $a_3'a_4'$. Thus if these two transformations are applied in order to $a_3x_3^2 + a_4x_4^2$, f becomes g where g has integral coefficients and determinant d/p^2, which is prime to p. Second $c_p(f) = -1$, in which case d is not a square since f is a zero form in $F(p)$ and hence $(-a_3'a_4' \mid p) = -1$ implies $(-a_1a_2 \mid p) = 1$. Then take $a_1x_1^2 + a_2x_2^2$ into pg where the determinant of g is a_1a_2 and f goes into pf' where the determinant of f' is d/p^2.

By the above process we can eliminate all square factors of the determinant of f, thus getting a form d of square-free determinant which is a zero form if and only if f is and a p-adic zero form if and only if f is. If the absolute value of the determinant is a square we may by this process reduce it to a form of determinant ± 1 which may be dealt with as in the case $n = 3$, thus completing our proof.

It remains to consider forms of square-free determinants. Now from property 9 of the Hilbert symbol together with the fact that if d is a square in any $F(p)$ then $c_p(f) = 1$ for that p, we see that for every prime factor p of $2d$ and $p = \infty$ we can determine a w_p so that $c_p(f) = (d, w_p)_p$ with w_p prime to p unless $p = 2$ and d is odd. Then choosing an integer of the same sign as w_∞, congruent to $w_p \pmod{p}$ for p odd and congruent to $2w_2 \pmod{16}$ we see that $c_p(f) = (d, w)_p$ for all primes p dividing $2d$ and for $p = \infty$. Then, by theorem 26 there is an integer W such that $c_p(f) = (d, W)_p$ for all primes p including $p = \infty$. For this W it is true that $c_p(Wf) = (d, W)_p c_p(f) = 1$ by property 3 of the Hasse symbol.

Consider the form $g = Wf - x_5^2$. Then Pall's invariant

$$\begin{aligned} k_p(g) &= c_p(g)(-1, -W^4 d)_p \\ &= c_p(Wf)(-1, -W^4 d)_p(-1, -W^4 d)_p = c_p(f)(d, W)_p = 1 \end{aligned}$$

for all p. Furthermore g is indefinite for Wf a negative form would imply d a square in $F(\infty)$ and $c_\infty(Wf) = c_\infty(f) = -1$ from section 13 and f would not be a zero form in $F(\infty)$. Then, by the proof of the first part of case $n = 5$ below (which is independent of this case) we see that g is a zero form and hence Wf represents a square number N^2. Hence $Wf = M^2$ has a solution in integers whose g.c.d. is 1 and M is a divisor of N. Thus by the corollary to lemma 6 Wf is equivalent to a form whose leading coefficient is M^2. Hence Wf is rationally congruent to a form $x^2 + h$ where h is a ternary form. But $1 = c_p(Wf) = c_p(h)$ for all p and hence by the case $n = 3$, h is a zero form and thus f is.

Fourth suppose $n = 5$. As above we may assume that f is in diagonal form with 1 the g.c.d. of every three coefficients and each coefficient is square-free; hence d is not divisible by any cube except ± 1. If two coefficients are even we can show as in the previous case that f is rationally congruent to a form whose determinant is $d/4$ and hence odd. If the last coefficient only is even we use the transformation

$$\begin{bmatrix} 2 & 1 & 0 & 0 & 0 \\ 0 & 1 & 0 & 0 & 0 \\ 0 & 0 & 2 & 1 & 0 \\ 0 & 0 & 0 & 1 & 0 \\ 0 & 0 & 0 & 0 & 1 \end{bmatrix}$$

Though in the last two instances the resulting form is no longer diagonal, by permuting the variables, we may consider it the sum of a ternary form in x_1, x_2, x_3 and a binary form in x_4 and x_5. Moreover mod any power of an odd prime it may be taken to be diagonal.

If an odd prime p but not p^2 divides d, write f in diagonal form (mod p^2) and assume the last coefficient divisible by p. Then the first four terms constitute in $F(p)$ a quaternary zero form with determinant prime to p which can by theorem 24 be taken into pg where g is a quaternary form of determinant p^2; hence f is so taken into pf' where $|f'| = pd$. Then one coefficient of f' (mod p^2) considered to be diagonal, is divisible by p^2 which may be absorbed into the variable or p divides two coefficients of f' (mod p^2). Thus we reduce consideration to d in absolute value a perfect square and odd.

If $k_p(f) = 1$ and $f \equiv a_1x_1^2 + a_2x_2^2 + a_3x_3^2 + a_4x_4^2 + a_5x_5^2$ (mod p^2) with a_4 and a_5 divisible by p, we see $k_p(f) =$

$(-a_4a_5/p^2 \mid p) = 1$ and by the same argument as in the case $n = 4$ we take f into a form whose determinant is prime to p. If $k_p(f) = 1$ for all primes p dividing d we can continue the procedure to reduce consideration to $d = \pm 1$ which is dealt with just as for $n = 3$. This completely disposes of the case $n = 4$.

Now we consider the general case of the quinary form. By the first paragraph of this case, we may take $f = g - h$ where g is a ternary form and h a binary form. Multiplying by -1 if necessary we may assume g and h represent a positive number in the field of reals since f is a zero form in that field. Moreover d may be considered odd and not divisible by the cube of any prime. We seek a non-zero number M represented by g and h, for the existence of such a number will prove our theorem. For any odd prime p dividing d choose N_p as follows: if $\mid g \mid$ is prime to p, take N_p any number represented by h and not divisible by p^2; if $\mid g \mid$ and $\mid h \mid$ are divisible by p but neither by p^2, take N_p a number prime to p represented by h; if $\mid h \mid$ is prime to p and $\mid g \mid$ divisible by p, take N_p a number prime to p represented by g. In all cases, corollary 14 shows that N_p is represented by g and h in $F(p)$. Since d is odd, g represents at least three and h at least two incongruent numbers (mod 8) and hence by corollary 9b there is an odd number N_2 represented by g and h in $F(2)$. Now choose $N \equiv N_p$ modulo an arbitrary but fixed power of p for all primes p dividing $2d$ with $N > 0$. We see that $c_p(h) = (-\mid h \mid, -N)_p$ for all primes p dividing $2 \mid h \mid$ and $p = \infty$. Hence by theorem 26 we may find an $M \equiv N$ (mod $16d^4$) such that $c_p(h) = (-\mid h \mid, -N)_p$ for all primes p. Hence M is represented by h in all $F(p)$. Furthermore, $M \equiv N$ (mod $16d^4$) implies $g = M$ is solvable in F(p) for all primes p dividing $2d$; moreover if p does not divide $2d$, $c_p(g) = 1$ and $-\mid g \mid M$ a square would imply $(-\mid g \mid, -M)_p = (-\mid g \mid, -\mid g \mid M)_p = 1$ which shows that M is represented by g in $F(p)$ for all primes. Then by corollary 27a below for $n = 2, 3$ (which depends only on the cases $n < 5$) we see that M is represented rationally by g and h. Hence our proof is complete.

In view of corollary 13 we have

COROLLARY 27a. *If f is a form with rational coefficients and N a rational number, then $f = N$ has a solution in rational numbers if and only if it has a solution in $F(p)$ for every prime p.*

From theorem 14 and its corollary we have

REMARK 1. In theorem 27 and its corollary 27a only the infinite prime and primes dividing $2d$ or $2dN$, respectively, need be considered.

From section 13, $c_\infty(f)$ and $|f|$ determine the index of f if it has less than four variables since the sign of $|f|$ determines the parity of i. Furthermore it may be seen that for $n < 4$, theorem 14 and its corollary hold for $p = \infty$. Hence property 2 of the Hasse symbol and property 10 of the Hilbert symbol with theorem 14 and its corollary show

REMARK 2. Any form f in two or three variables with rational coefficients is a zero form in all $F(p)$ if it is for all but one prime and any binary form with rational coefficients represents a rational number in all $F(p)$ if it represents it in $F(p)$ for all but one prime p.

That the latter part of this remark fails for ternary forms is shown by the fact that $x_1^2 + 2x_2^2 + 5x_3^2 = 15$ is solvable in $F(p)$ for all primes p including $p = \infty$ except the single prime $p = 5$.

In the consideration of zero forms there is no essential restriction involved in restricting the coefficients and solutions to be integers. Then, using corollary 14, we shall establish

COROLLARY 27b. *If $f = a_1x_1^2 + a_2x_2^2 + a_3x_3^2$ where the* *coefficients are square-free integers and prime in pairs, then* $f = 0$ *has a non-trivial solution if and only if both of the following conditions hold:*

1. *Not all coefficients have the same sign.*
2. *For every* i, j, k *a permutation of* 1, 2, 3 *it is true that* $-a_i a_j$ *is a quadratic residue of* a_k.

For suppose an odd prime p divides a_3, then $c_p(f) = (-a_1a_2, -a_2a_3)_p = (-a_1a_2 \mid p)$ and f is a zero form in $F(p)$ if and only if $(-a_1a_2 \mid p) = 1$, that is, $-a_1a_2$ is a quadratic residue of p. If a_3 is even, then $-a_1a_2$ must be a quadratic residue of 2 since it is odd. These considerations with the condition that not all the coefficients shall have the same sign assures us that except perhaps for $p = 2$, f is a zero form in all $F(p)$. Then from remark 2 above it is also a zero form in $F(2)$ and theorem 27 establishes our result.

We illustrate these two corollaries by the following examples

EXAMPLE 1. To find the numbers represented rationally by $f = (5/3)x_1^2 - 2x_1x_2 + 7x_2^2$. Now f is a positive form. Hence a rational number N is represented by f if and only if it is positive and, using remark 1, $c_p(f) = (-96, -N)_p$ for $p = 2, 3$ and $(-96, -N)_p = 1$ for all prime factors not 2 or 3 of the numerator and denominator of N. Now $c_2(f) = 1 = -c_3(f)$. Hence the requirement becomes $(-6, -N)_p = 1$ for $p = 2$ and all odd primes dividing the numerator or denominator of N except that $(-6, -N)_3 = -1$. In particular if N is a positive integer prime to 6 this reduces to

$N \equiv 1 \pmod 3$, $\equiv \pm 1 \pmod 8$ and -6 a quadratic residue of all prime factors of N occurring in N to an odd power.

EXAMPLE 2. Suppose $f = x_1^2 + 5x_2^2 - 2x_2x_3 - 6x_3^2$. Then $D_1 = 1$, $D_2 = 5$, $D_3 = -31$ and section 13 shows that $c_2(f) = -1$, $c_\infty(f) = 1$. Since the product of $c_p(f)$ over all primes p is 1, we have $c_{31}(f) = -1$. Hence, from corollary 14, N is represented in all $F(p)$ unless $31N$ is a square in $F(2)$ or $F(31)$. Hence all rational numbers N are rationally represented except numbers of the forms $4^k r$ with $r \equiv 7 \pmod 8$ and $31^{2k+1}s$ where $(s \mid 31) = 1$, k being an integer positive, negative or zero and r and s are rational numbers. Notice that N need not be positive.

EXAMPLE 3. The form $f = 3x_1^2 + 7x_2^2 - 19x_3^2$ is a zero form since $(-21 \mid 19) = (57 \mid 7) = (133 \mid 3) = 1$.

Theorems 14c and 27 may be shown to imply

COROLLARY 27c. *If* $f = a_1x_1^2 + a_2x_2^2 + a_3x_3^2 + a_4x_4^2$ *where the coefficients are square-free non-zero integers and no three have a factor in common, then* f *is a zero form if and only if the following three conditions hold*

1. *Not all coefficients have the same sign.*
2. *If* p *is an odd prime dividing two coefficients and for which* $(d/p^2 \mid p) = 1$, *then* $(-a_ia_j \mid p) = 1$ *where* a_i a_j *are the coefficients of* f *prime to* p.
3. *If* $d \equiv 1 \pmod 8$ *or* $d/4 \equiv 1 \pmod 8$ *then* $(-a_1a_2, -a_2a_3)_2 = 1$. *(For* $d \equiv 1 \pmod 8$ *the condition amounts to the requirement that exactly two of the coefficients are congruent to* 1 mod 4).

The last two corollaries are classical and could, by means of theorem 14, be stated for forms with cross-products. Another immediate consequence of theorem 27 is

COROLLARY 27d. *Every indefinite form in 5 or more variables (with non-zero determinant) is a zero form.*

Theorem 13 and the above shows that any zero form in the field of rationals is a universal form. Theorem 19 implies

COROLLARY 27e. *Any universal form in the field of rationals is a zero form in the same field unless*

$n = 4$ and, for some prime p, d is a square in $F(p)$ and $c_p(f) = -1$.

The form $x_1^2 + 2x_2^2 + 5x_3^2 - 10x_4^2$ is universal in the field of rational numbers but is not a zero form in that field.

22. The fundamental theorem on rational congruence. We prove the following theorem on rational congruence which is analogous to theorem 27.

THEOREM 28. *If f and g are two forms of non-zero determinant and with rational coefficients then they are rationally congruent if and only if they are congruent in $F(p)$ for every prime p including $p = \infty$.*

If f and g are rationally congruent they must be congruent in $F(p)$ for every p.

Then we suppose f and g have determinants d_f and d_g and are congruent in every $F(p)$. This implies that they have the same number of variables which we call n. If $n = 1$, $d_f = t_p^2 d_g$ for every p, where t_p is a p-adic number, implies that d_f/d_g is a square in all $F(p)$ and in particular for all prime factors of the numerators and denominators of d_f and d_g. Thus $d_f/d_g = \pm s^2$ for s a rational number while congruence in the field of reals requires the ambiguous sign to be $+$. Then $x = sy$ takes f into g and the forms are congruent.

Now assume the theorem for $n = k - 1$ and, by induction, prove it for $n = k$. Let N be some non-zero number represented by f. It must satisfy for f the conditions of corollary 14 for every prime p. Since the determinants of f and g differ only by a square factor in $F(p)$ and the Hasse invariants are the same for both forms it follows that g represents N in $F(p)$ for every p. Hence, by theorem 27, g represents N. By the corollary of lemma 6, f and g are congruent to forms $\Sigma\, a_{ij}x_ix_j$ and $\Sigma\, b_{ij}y_iy_j$ where $a_{11} = b_{11} = N$. Replacing x_1 by

$$x_1 - (a_{12}/N)x_2 - (a_{13}/N)x_3 - \cdots - (a_{1n}/N)x_n$$

and making a similar replacement for y_1, we take f and g into forms

$$N + f_1 \quad \text{and} \quad N + g_1$$

where f_1 and g_1 are forms in $x_2, \cdots, x_n$ and $y_2, \cdots, y_n$ respectively. Now $N \mid f_1 \mid \sim d_f$ and $N \mid g_1 \mid \sim d_g$ and property 4 of the Hasse symbol shows that the Hasse invariants of f_1 and g_1 are the same while the quotient of their determinants is a square in $F(p)$ for every p. Thus f_1 and g_1 are congruent in $F(p)$ for every p and are, by the hypothesis of our induction, rationally congruent. Thus f and g are rationally congruent.

23. **Representation of one form by another.** This follows much the same lines as section 17 in the previous chapter. We first need the theorem corresponding to theorem 16, which is

THEOREM 29. *Given a positive integer n, a non-zero integer d, a set of values $+1$ or -1 for $c_p(f)$ for all primes including $p = \infty$ and a non-negative integer i not greater than n, there is a form with n variables, integer coefficients, determinant d, Hasse invariants of the given values and index i if and only if the following conditions hold:*

1. *$c_p(f) = +1$ for finite primes p not dividing $2d$.*
2. *$\Pi c_p(f) = +1$, the product extending over all primes including $p = \infty$.*
3. *If $n = 1$, $c_p(f) = (-1, -d)_p$ for all p including $p = \infty$.*
4. *If $n = 2$, then $c_p(f) = 1$ for every prime p, including $p = \infty$, for which $-d$ is a square.*
5. *$n - i \equiv c_\infty(f) + \frac{1}{2}\{1+(-1, d)_\infty\} \pmod 4$.*

Theorem 16 shows that conditions 3 and 4 are necessary,

properties 1 and 2 of the Hasse symbol that conditions 1 and 2 are necessary and (9b) of section 13 that condition 5 is necessary. If $n < 4$ condition 5 determines the index and hence for such values of n we need not consider the index.

If $n = 1$, d determines $c_p(f)$ by condition 3 and the form dx^2 has the required invariants.

If $n = 2$ we may, by property 9 of the Hilbert symbol, choose a square-free integer a such that $(-a, -d)_p$ has the required value for $c_p(f)$ for $p = \infty$ and every prime p dividing $2d$. Furthermore if w is the g.c.d. of a and $2d$ we may assume that each prime factor of w and hence w itself occurs in d to an even power. Hence if we write d in the form $d_0 s^2$ where d_0 is square-free, w divides $2s$ and is prime to d_0. Then, by Dirichlet's theorem on the primes in an arithmetic progression (which can be avoided by composition of binary forms) there is a prime $P \equiv a/w \pmod{8d^2}$ since a/w is prime to $2d$. Then $(-wP, -d)_p = (-a, -d)_p$ for all primes p dividing $2d$ and, giving P the sign of a, we have $\Pi(-wP, -d)_p = 1$ the product being over all primes except P. Hence $(-wP, -d)_P = 1 = (-wP, -d_0)_P$, that is, $(-d_0 \mid P) = 1$ which implies that the equation $-d_0 = u^2 - Pv$ is solvable for integers u and v. Then the form

$$f = Pwx^2 - 2suxy + (s^2v/w)y^2$$

has determinant d and the required values for its invariants.

If $n = 3$, let w be the product of primes dividing $2d$ for which $c_p(f) = -1$, determine by Dirichlet's theorem a prime Q not dividing $2d$ such that $(-wQ, d)_p = 1$ for all prime factors p of $2d$ not in w. Then find a binary form $g = ax^2 + 2txy + by^2$ of determinant $+wQ$ such that $c_p(g) = (-wQ, d)_p c_p(f)$ for $p = \infty$ and all primes p

dividing $2dQ$. This is possible by the argument in the previous paragraph since $-wQ$ a square in $R(p)$ for some p implies that p divides neither w nor Q and p finite; but for such primes $c_p(f) = 1 = c_p(g)$. Since $a \equiv b \equiv 0 \pmod{Q}$ would deny $|g| = -wQ$, a may be taken prime to Q. Then since a is prime to Q and $c_Q(f) = 1$, $c_Q(g) = (-a, -wQ)_Q = (-wQ, d)_Q c_Q(f)$ which implies that $(-ad, -wQ)_Q = 1$ and $(-ad \mid Q) = 1$. Hence $-d \equiv ax^2 \pmod{Q}$ is solvable, $-d/w \equiv awx^2 \pmod{Q}$ is solvable, $-d \equiv ax^2 \pmod{wQ}$ is solvable and hence there are integers c and r such that $d = cwQ - ar^2$. Then the integral form

$$f = ax^2 + 2txy + by^2 + cz^2 + 2ryz$$

has the required invariants by the properties of the Hasse invariants.

If $n > 3$ we consider two cases. If $i > 0$ we may by induction find an integral form f' of determinant d, index $i - 1$, having $n - 1$ variables (since the satisfaction of condition 5 for n and i is equivalent to its satisfaction for $n - 1$ and $i - 1$) such that $c_p(f')$ has the chosen value of $c_p(f)$ for all primes p including $p = \infty$. Then $f' + x^2$ has the required invariants. If $i = 0$ there is, by induction, an integral form f' in $n - 1$ variables such that $c_p(f' - x^2)$ has the given values for all p including $p = \infty$. To show that the index of f' may be taken to be 0 we use the fact that condition 5 holds for f with $i = 0$. An equivalent form of this condition is, by section 13,

$$(14) \qquad d(-1)^n > 0, \qquad c_\infty(f) = (-1)^{(n-1)(n-2)/2}.$$

But, by property $4'$ of the Hasse symbol, $c_\infty(f' - x^2) = c_\infty(f')\,(d, -1)_\infty$. Hence (14) implies

$$(15) \qquad \begin{aligned} &|f'|\,(-1)^{n-1} > 0, \\ &c_\infty(f') = (-1)^{n+(n-1)(n-2)/2} = (-1)^{(n-2)(n-3)/2} \end{aligned}$$

since $2n + (n - 1)(n - 2) \equiv (n - 2)(n - 3) \pmod 4$ and condition 5 holds for f' with $i = 0$.

An immediate consequence of this theorem is

COROLLARY 29. *If f is an n-ary quadratic form with integral coefficients and index i, then for any non-negative integer i' not greater than n and congruent to i (mod 4) there is an n-ary quadratic form f' with integral coefficients and index i' such that $|f'| = |f|$ and $c_p(f) = c_p(f')$ for all primes p including $p = \infty$.*

If f is a form with rational coefficients the l.c.m. of whose denominators is s, we may consider the form sf in place of f and apply the above theorem.

As in the previous chapter, this theorem could be used to construct a unique canonical form, but it seems scarcely worth doing.

We can now prove

THEOREM 30. *If f and g are forms with rational coefficients in n_f and n_g variables, having indices i_f and i_g and determinants d_f and d_g, where $n_f > n_g > 0$, then f represents g in the field of rational numbers if and only if the following conditions hold:*

1. $i_f \geqq i_g$, $n_f - i_f \geqq n_g - i_g$.
2. *If $n_f - n_g = 1$, then $c_p(f)c_p(g) = (-d_g, d_f)_p$ for all primes p.*
3. *If $n_f - n_g = 2$, then $c_p(f)c_p(g) = (-1, -d_g)_p$ for every prime p for which $-d_f d_g$ is a square.*

Theorem 5 applies to prove that f represents g if and only if there is a form h such that f is rationally congruent to $g + h$. Now property 4 of the Hasse symbol and theorems 12, 15 and 28 show that f is rationally congruent to $g + h$ if and only if $d_f \sim d_h d_g$ and

$$c_p(h) = c_p(f)c_p(g)(-1, -1)_p(d_g, d_f d_g)_p, \text{ for all } p. \tag{16}$$

Hence we determine the conditions that such an h will exist. The conditions of this theorem are necessary from theorem 6 and conditions 4 and 3 of theorem 29. To prove the conditions sufficient let c be an integer such that cd_f/d_g is an integer. If each variable of f is multiplied by c, f is replaced by a form f' congruent to it and $|f'|/d_g$ is an integer. Hence we may take d_f/d_g to be an integer. Thus our theorem will be proved if we can find a form h in $n_f - n_g$ variables, having integer coefficients, the integral determinant d_f/d_g and $c_p(h)$ determined by (16). But (16) and properties 1 and 2 of the Hasse symbol assure the satisfaction of conditions 1 and 2 of theorem 29; conditions 2 and 3 of this theorem with (16) assure the satisfaction of conditions 3 and 4 of theorem 29; condition 5 of theorem 29 need not be considered since the index of h is not preassigned. Hence our proof is complete.

Then theorems 6, 17, 28, 30 imply

THEOREM 31. *If f and g are two forms with rational coefficients and if the number of variables in f is not less than the number of variables in g, then f represents g rationally if and only if it represents g in $F(p)$ for every prime p.*

Notice that theorem 7 holds for rational congruence. Theorem 31 taken with theorems 18 and 21 give conditions that a form be an m-zero or m-universal form in the field of rational numbers.

Chapter IV

FORMS WITH COEFFICIENTS IN $R(p)$

24. **Equivalence.** We further restrict our forms in this chapter by requiring that their coefficients be p-adic integers, that is, in $R(p)$. As the theory for $F(p)$ led to results for forms with rational coefficients, so the theory for $R(p)$ leads to results for forms with integer coefficients though, as we shall see, there is here an intermediate case which has no previous analogy. Suppose a transformation T with elements in $R(p)$ takes a form f into a form g where both forms are in $R(p)$ and have non-zero determinants d_f and d_g, respectively, and a transformation S in $R(p)$ takes g into f. Then

$$| T^2 | \cdot d_f = d_g \quad \text{and} \quad | S^2 | \cdot d_g = d_f$$

imply that $| T |^2 | S |^2 = 1$. Since the determinants of T and S are p-adic integers, they must be units. Thus, following section 20, we call a transformation *unimodular* in $R(p)$ or *p-adically unimodular* if its elements are in $R(p)$ and its determinant a p-adic unit. Two forms are *equivalent* in $R(p)$ or *p-adically equivalent* if one may be taken into the other by a unimodular transformation in $R(p)$. Two p-adically equivalent forms represent the same p-adic integers for values of the variables in $R(p)$. In this chapter, unless it is stated to the contrary, equivalence means p-adic equivalence and we denote it by the sign $\cong$. Two forms equivalent in $R(p)$ are said to be in the same *p-adic class.*

25. **Canonical forms.** Here we begin to have trouble for the prime 2. (The reason for this trouble is not any fundamental perversity of the prime 2 but is due to

the fact that quadratic forms are the subject of our study. If it were cubic forms, 3 would be the onerous prime.) But for odd primes we use the proof of theorem 25 with the following alteration: in the second paragraph we can choose the r's to be p-adic integers satisfying the equation $(a_{11})x + a_{1i} = 0$. Thus, using the argument of section 9, we have

THEOREM 32. *If f is a form with coefficients in $R(p)$ where p is an odd prime, then f is (p-adically) equivalent to a form*

$$g = a_1x_1^2 + a_2x_2^2 + \cdots + a_nx_n^2$$

where the coefficients are integers.

That this theorem does not hold for $p = 2$ is shown by taking as f the form $2x_1^2 + 2x_1x_2 + 2x_2^2$. If $g = a_1x_1^2 + a_2x_2^2$ were equivalent to f its determinant would have to be a unit and hence a_1 and a_2 would be units. Then g would represent a unit while f represents no units. Thus g cannot be p-adically equivalent to f. To distinguish between two such forms the following terminology is useful. The form $f = \Sigma\, a_{ij}x_ix_j$ in $R(2)$ is called *properly primitive* if some a_{ii} is a unit and *improperly primitive* if no a_{ii} is a unit but some a_{ij}, $i \neq j$ is a unit. Notice that any improperly primitive form f may be written in the form $f = 2g$ where g has coefficients in $R(2)$. In this case the matrix of g will contain at least one element not on the principal diagonal which is half a unit though all its elements on the principal diagonal are 2-adic integers. Such a form g is called *non-classic*, that is, g is non-classic if $2g$ is improperly primitive. Forms whose matrices are in $R(2)$ are *classic*. In $R(p)$, however, for p odd all forms are classic and none are improperly primitive. For p odd or $p = 2$ a form is called *primitive* if 1 is the g.c.d. of the elements of its natrix. Unless something is written to the contrary we deal only with classic forms. There is no loss of generality in so doing.

Theorem 32 implies the following theorem for p odd. For $p = 2$ we give the proof.

THEOREM 33. *If f is a form with coefficients in $R(p)$, it is (p-adically) equivalent to a form*

$$g = p^{t_1}g_1 + p^{t_2}g_2 + \cdots + p^{t_r}g_r$$

where $0 \leqq t_1 < t_2 < \cdots < t_r$ and the g_i are classic forms of unit determinant in $R(p)$. Furthermore, each g_i may be considered to be diagonal or, when $p = 2$, the sum of improperly primitive binary forms.

Let 2^t be the greatest power of 2 dividing all the elements of the matrix of g and write $g = g_0 2^t$ where g_0 is a classic form having at least one unit element in its matrix. If some element in the principal diagonal of the matrix of g_0 is a unit, permute variables to make it the leading coefficient and add appropriate multiples of the first column of the matrix to the other columns and similarly for the rows to make the first row and column $a_{11}, 0, 0, \cdots, 0$. If all diagonal elements of the matrix are non-units we may without loss of generality take a_{12} a unit and see that $a_{11}a_{22} - a_{12}^2$ is a unit and appropriate multiples of the first two columns of the matrix may be added to the other columns and similarly to the rows to make all the elements in the first two rows and columns 0 except the leading two by two principal minor. This process may be continued to show that g is expressible in the given form where each g_i is a sum of unary and improperly primitive binary forms.

It remains to show that if g is a form in $R(2)$ of unit determinant it is equivalent to a diagonal form or a sum of improperly primitive binary forms, that is, if g is properly primitive and has unit determinant, it is equivalent to a diagonal form. To this end notice that one diagonal element of its matrix must be a unit if g represents a unit and the transformation

$$\begin{bmatrix} 1 & 1 & 1 \\ 1 & 1 & 0 \\ 1 & 0 & 1 \end{bmatrix}$$

takes $a_{11}x_1^2 + a_{22}x_2^2 + 2a_{23}x_2x_3 + a_{33}x_3^2$ into

$$h \equiv x_1^2(a_{11} + a_{22} + 2a_{23} + a_{33}) + x_2^2(a_{11} + a_{22}) + x_3^2(a_{11} + a_{33}) \pmod 4$$

if a_{22} and a_{33} are non-units and a_{23} a unit. Then by adding even multiples of the first two columns of h to the other columns and similarly for the rows, we can reduce h to diagonal form with units along the diagonal. Continuing this process we can get the desired result.

This completes the proof of theorem 33 and proves part of

THEOREM 33a. *Every form in $R(2)$ with unit determinant is equivalent to one of the following:*

1. $a_1x_1^2 + a_2x_2^2 + \cdots + a_nx_n^2$, *$a_i$ odd integers.*
2. *A sum of binary forms of the two types:* $f = 2x_1^2 + 2x_1x_2 + 2x_2^2$, $g = 2x_1x_2$.

Furthermore in the second case one may at will choose one of the types f and g and require that all but at most one of the binary forms be of that chosen type. This we call a canonical form for forms of unit determinant.

From the previous theory we need only consider the improperly primitive case. Now if a_{11} and a_{22} are even and a_{12} is odd $h = a_{11}x_1^2 + 2a_{12}x_1x_2 + a_{22}x_2^2 \equiv 0 \pmod 8$ has a solution r_1, r_2 at least one of which is odd unless $a_{11} \equiv a_{22} \equiv 2 \pmod 4$. Thus, barring the exception, corollary 9b and the corollary of lemma 6 show that $h \cong 2a'_{12}x_1x_2 + a'_{22}x_2^2$ with a'_{22} even which is taken into $2x'_1x'_2$ by the transformation $2x'_1 = 2a'_{12}x_1 + a'_{22}x_2$, $x'_2 = x_2$.

In the exceptional case one of $a_{11} \pm 2a_{12} + a_{22}$ is $\equiv 2 \pmod 8$ showing that $h \cong 2x_1^2 + 2a'_{12}x_1x_2 + a'_{22}x_2^2$, where $a'_{22} \equiv 2 \pmod 4$. Then if we let α be that 2-adic integer for which $3 = (2a'_{22} - a'^2_{12})\alpha^2$ we see that the transformation $x_1 = y_1 + \frac{1}{2}(1 - \alpha a'_{12})y_2$, $x_2 = \alpha y_2$ takes the last form into $2y_1^2 + 2y_1y_2 + 2y_2^2$.

The last part of the theorem is shown by the fact that the transformation

$$\begin{pmatrix} 1 & 1 & 1 & 0 \\ 1 & 1 & 0 & 1 \\ 1 & 0 & 1 & 1 \\ 0 & -1 & -1 & -1 \end{pmatrix}$$

takes $x_1x_2 + x_3x_4$ into $x_1^2 + x_2^2 + x_1x_2 - x_3^2 - x_4^2 - x_3x_4$. Hence, since we have shown above that $-(x_3^2 + x_4^2 + x_3x_4) \cong x_3^2 + x_4^2 + x_3x_4$, we have $2(x_1x_2 + x_3x_4) \cong 2x_1^2 + 2x_1x_2 + 2x_2^2 + 2x_3^2 + 2x_4^2 + 2x_3x_4$. This means that we can, in the canonical form, replace any two of one of the types of binary improperly primitive forms by two of the other type.

26. **Representation of numbers by forms.** While a zero form in $F(p)$ is a zero form in $R(p)$, the solvability of $f = N$ in $F(p)$ does not imply its solvability in $R(p)$ as is shown by the fact that $x_1^2 + 9x_2^2$ represents 2 in $F(3)$ but not in $R(3)$. In general the best method for testing for the solvability of $f = N$ in $R(p)$ is by use of theorem 9a or 9b. However, under certain restrictions we can obtain more specific results.

Then suppose p is an odd prime and f is a form with coefficients in $R(p)$ and of unit determinant. Then $f = N$ solvable in $F(p)$ is equivalent to saying that, for some k, $f = Np^{2k}$ is solvable in $R(p)$. Let $r_1, r_2, \cdots, r_n$ be a solution. If all r_i are divisible by p we may divide them and the right side of the equation by p and p^2 respectively to have a "primitive" solution of $f = Np^{2t}$ where N is not divisible by p^2. (By a solution being "primitive" we mean that not all the r's are divisible by p.) If $t = 0$ we have a solution of $f = N$ in $R(p)$. If $t > 0$, corollary 9b shows that $f = 0$ has a non-trivial solution in $R(p)$ and hence by theorem 13 and the note thereafter, f is a universal form in $R(p)$ and hence represents N. Thus we have

THEOREM 34. *If f is a form of unit determinant in $R(p)$ where p is odd and N a number in $R(p)$, then f represents N in $R(p)$ if and only if it represents N in $F(p)$.*

Remark. It may similarly be shown that the theorem holds if $f = f_1 + pf_2$ where f_1 and f_2 are forms in $R(p)$ of unit determinant.

This theorem with corollary 14 gives us the following interesting results.

COROLLARY 34a. *If f is a binary form in $R(p)$ of non-zero determinant and p does not divide $2d$, then $f = N$ is solvable in $R(p)$ if and only if either p occurs to an even power in N or p occurs to an odd power with $(-d \mid p) = 1$.*

COROLLARY 34b. *If f is a ternary form in $R(p)$ of non-zero determinant d and p does not divide $2d$, then $f = N$ is solvable in $R(p)$.*

The theorem does not hold in $R(2)$ as is shown by the fact that $x_1^2 + 7x_2^2$ represents 2 in $F(2)$ but not in $R(2)$. However we now show that the theorem holds even for this case with essentially two exceptions. Then let f be a properly primitive form of unit determinant in $R(2)$. As in the proof for p odd we see that $f = N$ is solvable in $F(2)$ if and only if, for some integer k, $f = 4^kN$ is solvable in $R(2)$ where N is an integer not divisible by 4. We may assume the solution in $R(2)$ to be primitive, that is, that one of the variables assumes a unit value. If $k > 1$, corollary 9b shows that $f = 0$ has a non-trivial solution in $R(2)$ and hence by the proof of theorem 13, $f = 4N$ has a solution in $R(2)$. If the solution is non-primitive, then $f = N$ is solvable and our result follows. Furthermore, trial shows that $f \equiv a \pmod 8$ is solvable primitively for all $a \not\equiv 0 \pmod 4$ and hence by corollary 9b f is universal unless f is a ternary form with all coefficients congruent (mod 4) or f is a binary form. In the former exceptional case $f = 4N$ has no primitive solution. In the latter it has no primitive solution unless $f = a_1x_1^2 + a_2x_2^2$ with $a_1a_2 \equiv 3$ (mod 4). If N is odd it is then represented by f in $R(2)$. If $N \equiv 2 \pmod 4$, $f = 4N$ solvable primitively implies $a_1a_2 \cong -1$ in $F(2)$ and N is then not represented by f in $R(2)$.

If f is improperly primitive, notice that $2x_1x_2$ is universal in $F(2)$ and represents all non-units in $R(2)$. Hence theorem 33a shows that all improperly primitive forms are universal in $F(2)$ except $f \cong 2x_1^2 + 2x_1x_2 + 2x_2^2$ which in $F(2)$ and $R(2)$ represents all numbers not a unit multiple of a power of 4. Hence we have

THEOREM 34a. *If f is a form of unit determinant in $R(2)$ and N a number in $R(2)$, then $f = N$ is solvable in $R(2)$ if and only if it is solvable in $F(2)$ with the following exceptions:*

1. *N a unit and f an improperly primitive form* $\not\cong 2x_1^2 + 2x_1x_2 + 2x_2^2$.
2. *N twice a unit and* $f \cong a_1x_1^2 + a_2x_2^2$ *with* $a_1a_2 \cong -1$.

27. **Equivalence of forms in** $R(p)$. Since for p-adic integers we have no such clear-cut criteria for representation of a number by a form as in the previous chapters, our task of finding conditions for equivalence is more complex though our criteria for equivalence turn out to be relatively simple. We must expect for $R(p)$ invariants in addition

to those involved in $F(p)$ and it will be in terms of these new invariants with the old ones that the criteria for equivalence will be phrased.

We first prove that, among other things, the exponents in each g_i in the form of theorem 33 are invariants under unimodular transformations.

THEOREM 35. *If f and g are p-adically equivalent forms with coefficients in $R(p)$ and are written*

$$f = p^{t_1}f_1 + p^{t_2}f_2 + \cdots + p^{t_j}f_j,$$

$$g = p^{s_1}g_1 + p^{s_2}g_2 + \cdots + p^{s_k}g_k,$$

with $t_i < t_{i+1}$ and $s_i < s_{i+1}$ for every i and all f_i and g_i of unit determinant then $k = j$ and, for every i, $s_i = t_i$, the number of variables in f_i is the same as the number of variables in g_i and, for $p = 2$, f_i and g_i are both improperly primitive or both properly primitive.

Before proving this theorem we need a lemma analogous to lemma 5.

LEMMA 7. *Let A be a symmetric matrix (α_{ij}) with elements in $R(2)$ and T a unimodular matrix in $R(2)$. Let $\mu_r(A)$ be the* g.c.d. *of the following set of minor determinants of A: r-rowed principal minors, the doubles of the r-rowed non-principal minors. Then $\mu_r(A) = \mu_r(T^TAT)$.*

As in the proof of lemma 5, we need consider T of one of three types. If T permutes the columns of A or multiplies the columns by units it is clear that $\mu_r(A) = \mu_r(T^TAT)$. Let T be a transformation E_2 which adds an integral multiple of one row to another row of the identity matrix. Then, if A_r is some r-rowed principal minor $E_2^TA_rE_2$ is obtained from A_r by performing that operation on the columns and rows of A_r. If the operation described is within A_r, its determinant is left unaltered. It then remains to consider the following typical case: Let D_r be the principal r-rowed minor in the upper left hand corner of A and let E_2 have the effect of adding α times the $r + 1$ row and column to the r-th row and column of D_r, where α is in $R(2)$. Then let α_1 and α_2 be the row matrices:

$$(\alpha_{r1}, \alpha_{r2}, \cdots, \alpha_{r,r-1}), \qquad (\alpha_{r+1,1}, \alpha_{r+1,2}, \cdots \alpha_{r+1,r-1})$$

respectively and D_{r-1} the matrix (α_{ij}) with $i, j \leqq r - 1$. Then

$$E_2^T D_r E_2 = \begin{bmatrix} D_{r-1} & \alpha_1^T + \alpha\alpha_2^T \\ \alpha_1 + \alpha\alpha_2 & \alpha_{rr} + 2\alpha\alpha_{r,r+1} + \alpha^2\alpha_{r+1,r+1} \end{bmatrix}$$

whose determinant is

$$|D_r| + 2\alpha \begin{vmatrix} D_{r-1} & \alpha_1^T \\ \alpha_2 & \alpha_{r,r+1} \end{vmatrix} + \alpha^2 \begin{vmatrix} D_{r-1} & \alpha_2^T \\ \alpha_2 & \alpha_{r+1,r+1} \end{vmatrix}.$$

Since the first and third determinants of this sum are r-rowed principal minors of A this shows that $\mu_r(A)$ divides $\mu_r(E_2^T A E_2)$. Similarly, the latter divides the former and hence they are equal.

Now, reverting to the proof of the theorem and noticing lemma 5 we have, using the notation of that lemma, $\lambda_1(A) = \lambda_1(B)$ where A and B are the matrices of f and g. Hence $p^{t_1} = p^{s_1}$ which implies $t_1 = s_1$; furthermore $m_1 = n_1$ where m_i and n_i are the number of variables in f_i and g_i respectively, since $\lambda_{m_1}(A/p^{t_1}) = \lambda_{m_1}(B/p^{s_1})$ implies that both are units. Then

$$\lambda_{m_1+1}\ (A/p^{t_1}) = p^{t_2-t_1}, \lambda_{n_1+1}(B/p^{t_1}) = p^{s_2-s_1}$$

implies that $s_2 = t_2$ while

$$\lambda_{m_1+m_2}(A/p^{t_1}) = p^{(t_2-t_1)m_2} = \lambda_{m_1+m_2}(B/p^{t_1})$$

implies that $m_2 = n_2$. This argument can be continued to show that $s_i = t_i$ and $m_i = n_i$ for all i and hence that $k = j$.

If $p = 2$, use lemma 7 to see that $\mu_1(A) = 2^{t_1}$ or 2^{t_1+1} according as f is properly or improperly primitive. Thus $\mu_1(A) = \mu_1(B)$ implies that f_1 and g_1 are both properly or improperly primitive. Then $\mu_{m_1+1}(A/2^{t_1}) = 2^{t_2} \cdot \mu_{m_1}(A/2^{t_1})$ or $2^{t_2+1} \cdot \mu_{m_1}(A/2^{t_1})$ according as f_2 is properly or improperly primitive. This shows that f_2 and g_2 are both properly or both improperly primitive. So we may proceed to prove the theorem.

It is not hard to prove

THEOREM 36. *If f and g are two forms of unit determinant in $R(p)$ they are equivalent in $R(p)$ if and only if they are congruent in $F(p)$ unless $p = 2$, one form is properly primitive and the other improperly primitive. Furthermore if $p = 2$ and f and g are improperly primitive they are equivalent if and only if their determinants are equivalent and they have the same number of variables.*

The last sentence ot our theorem follows directly from theorem 33a. It therefore remains to consider f and g diagonal forms $\Sigma\, \alpha_i x_i^2$ and $\Sigma\, \beta_i y_i^2$.

The truth of the theorem is obvious if $n = 1$. Assume it holds for $n < k$. Now $f \sim g$ in $F(p)$ implies that $f = \beta_1$ is solvable in $F(p)$ and hence, since β_1 is prime to p, is solvable in $R(p)$. Then, by permuting variables we may assume that $r_1, r_2, \cdots, r_k$ is such a solution with r_1 a unit. Let T be the transformation whose first column is the solution, whose first row is zero except for the first element and the rest of which is the $k - 1$ rowed identity matrix. Then T takes f into a form f_0 in $y_1, \cdots, y_k$ whose leading coefficient is β_1 and for which the coefficient of y_2^2 is prime to p. In fact

$$f_0 = \beta_1 x_1^2 + \sum_{i=2}^{k} 2\alpha_i r_i x_1 x_i + \sum_{i=2}^{k} \alpha_i x_i^2$$

$$= \beta_1 \left(x_1 + \beta_1^{-1} \sum_{i=2}^{k} \alpha_i r_i x_i\right)^2 + \sum_{i=2}^{k} (\alpha_i - \beta_1^{-1} \alpha_i^2 r_i^2) x_i^2$$

$$+ 2 \sum_{2=i<j}^{k} \gamma_{ij} x_i x_j = \beta_1 x_1'^2 + f_1,$$

for certain γ_{ij} in $R(p)$. Now if $p = 2$, f_1 will represent a unit unless all r_i are units. Postponing this exceptional case, we have $g \sim \beta_1 x_1^2 + f_1$ implies, by theorem 8, that $f_1 \sim g_1$ in $F(p)$ where $g_1 = \beta_2 y_2^2 + \cdots + \beta_k y_k^2$. Since f_1 and g_1 represent units, our induction hypothesis shows that $f_1 \cong g_1$ in $R(p)$, hence $\beta_1 x_1^2 + f_1 \cong g$ and $f \cong g$. To eliminate the exceptional case notice that $n \leqq 3$, $\beta_i \equiv -\alpha_j \pmod 4$ for all i, j would make $|f| \not\cong |g|$ or $c_2(f) \neq c_2(g)$ and hence would deny $f \sim g$ in $F(2)$; hence the congruence $f \equiv \beta_1 \pmod 4$ and hence $f \equiv \beta_1 \pmod 8$ has a solution with one variable a non-unit and by theorem 9a we may choose r_i in $R(2)$ such that one is a non-unit and $f \sim \beta_1 x_1^2 + f_1$ where f_1 has a unit coefficient.

If p is odd we know that the Hasse invariant of a form of unit determinant in $R(p)$ is 1. If $p = 2$, the Hasse invariant of such a form depends on the canonical form (mod 4) and the determinant depends on the form (mod 8). Thus theorems 12, 15 and 36 imply the following corollaries

Corollary 36a. *If f and g are two forms of unit determinant in $R(p)$ they are equivalent in $R(p)$ if $f \equiv g \pmod{4p}$. (The congruence of forms means congruence of corresponding coefficients. It includes the condition that the number of variables in each be the same.)*

Corollary 36b. *If p is odd, two forms of unit determinant in $R(p)$ are congruent if and only if they have the same number of variables and $|f| \cong |g|$. Every form f of unit determinant is equivalent to $x_1^2 + x_2^2 + \cdots + |f| x_n^2$.*

Corollary 36c. *Two properly primitive forms of unit determinant in $R(2)$ are equivalent if they are congruent* (mod 4) *and their determinants are congruent* (mod 8). *Two improperly primitive forms of unit determinant in $R(2)$ are equivalent if they are congruent* (mod 4), *their determinants are equivalent, or their Hasse invariants equal; the last two conditions are also necessary.*

In order to establish criteria for equivalence in the general case we need the following theorem proved by the author and W. H. Durfee. In the interests of brevity we state it for all primes p though much of the proof is concerned only with $p = 2$. Notice that some of the conditions are vacuous if p is odd.

THEOREM 37. *Let f_1 and f_2 be two equivalent forms of unit determinant in $R(p)$ with variables $x_1, x_2, \cdots, x_n$, g a form in $x_{n+1}, \cdots, x_{n+m}$ and h a form in $x_{n+1}, \cdots, x_{n+s}$, all forms being in $R(p)$; then if $f_1 + 2g$ represents $f_2 + 2h$ in $R(p)$, it follows that g represents h in $R(p)$. If $m = s$, the word "represents" may be replaced by "is equivalent to".*

Notice first that $s \leqq m$ and that $f_1 \cong f_2$ and the conditions of the theorem imply that $f_2 + 2g$ represents or is equivalent to $f_2 + 2h$. Thus we set $f_1 = f_2 = f$. By theorems 32 and 33a f is equivalent to either a diagonal form or, when $p = 2$, a sum of binary forms. If the latter is the case, then $x^2 + f$ will be equivalent to a diagonal form and since $x^2 + f + 2g$ represents or is equivalent to $x^2 + f + 2h$ we may assume that f is a diagonal form of unit determinant. Let f, g and h have respectively F, G and H as their matrices and let Q be the matrix taking $F \dotplus 2G$ into $F \dotplus 2H$. Write

$$Q = \begin{bmatrix} T & S_2 \\ S_3 & S_4 \end{bmatrix}$$

where T is an n by n matrix, S_2 is n by s, S_3 is m by n and S_4 is m by s. Then $Q^T(F \dotplus 2G)Q = F \dotplus 2H$ yields the following equations:

$$T^TFT + S_3^T 2GS_3 = F, \tag{17}$$

$$T^TFS_2 + S_3^T 2GS_4 = 0, \tag{18}$$

$$S_2^TFS_2 + S_4^T 2GS_4 = 2H. \tag{19}$$

We show in the following lemma that, since equation

(17) implies $T^T F T \equiv T \pmod 2$, there is an automorph D of F over $R(p)$ such that $2(T + D)^I$ is in $R(p)$. Furthermore, equation (18) and the fact that, for $p = 2$, $|T^T F|$ is a unit implies $S_2 \equiv 0 \pmod 2$ and hence that $S = S_4 - S_3(T + D)^I S_2$ is in $R(p)$. We shall show that $S^T 2GS = 2H$. Now

$$S^T 2GS = [S_4^T - S_2^T (T + D)^{TI} S_3^T] 2G [S_4 - S_3(T + D)^I S_2]$$
$$= S_4^T 2GS_4 - S_2^T (T + D)^{TI} S_3^T 2GS_4$$
$$- S_4^T 2GS_3(T + D)^I S_2 + S_2^T (T + D)^{TI}$$
$$\cdot S_3^T 2GS_3(T + D)^I S_2 .$$

Using (17) we have

$$S_2^T (T + D)^{TI} S_3^T 2GS_3 (T + D)^I S_2$$
$$= S_2^T (T + D)^{TI} \{F - T^T F T\} (T + D)^I S_2 . \tag{20}$$

Since $F - T^T F T = (D + T)^T F (D + T) - T^T F (D + T) - (T + D)^T F T$, the right side of (20) is equal to

$$S_2^T F S_2 - S_2^T (T + D)^{TI} T^T F S_2 - S_2^T F T (T + D)^I S_2 ,$$

which, using (18), becomes

$$S_2^T F S_2 + S_2^T (T + D)^{TI} S_3^T 2GS_4 + S_4^T 2GS_3 (T + D)^I S_2 .$$

Hence $S^T 2GS = S_4^T 2GS_4 + S_2^T F S_2 = 2H$ from equation (19). Thus we have shown that g represents h in $R(p)$.

It remains to prove the lemma.

LEMMA 8. *If F is a diagonal matrix of unit determinant in $R(p)$ and T a matrix in $R(p)$ such that $T^T F T \equiv F$* (mod 2), *then there is an automorph D of F in $R(p)$ such that $2(T + D)^I$ is in $R(p)$. (Notice that the congruence condition is vacuous unless $p = 2$.)*

Let $T = (t_{ij})$, $i, j = 1, 2, \cdots, n$ and $F = a_1 \dotplus a_2 \dotplus \cdots \dotplus a_n$. Since permuting the rows and the same columns of

T and F does not alter the properties we desire, we shall do this at will. Note that $T^T T \equiv I \pmod 2$.

First suppose that p is odd or that $p = 2$ with t_{ii} a non-unit for some i. Permute rows and columns, if necessary, to make t_{11} a non-unit when $p = 2$. Then choose $D = \pm 1 \dotplus D_1$ and $\mu = t_{11} \pm 1$ where the sign is so chosen that μ is a unit. If we set

$$P = \begin{bmatrix} 1 & -\mu^{-1}T_2 \\ 0 & I \end{bmatrix} \quad \text{and} \quad T = \begin{bmatrix} t_{11} & T_2 \\ T_3 & T_4 \end{bmatrix}$$

where T_2 is a 1 by $n - 1$ matrix, T_3 is $n - 1$ by 1 and T_4 is $n - 1$ by $n - 1$, we have

$$(21) \qquad (T + D)P = \begin{bmatrix} t_{11} \pm 1 & 0 \\ T_3 & T_1 + D_1 \end{bmatrix}$$

where $T_1 = T_4 - T_3\mu^{-1}T_2$ and D_1 is to be chosen an automorph of $F_1 = a_2 \dotplus \cdots \dotplus a_n$. By adding appropriate multiples of the first row of (21) to the later rows we can replace T_3 by 0 without altering $T_1 + D_1$. Referring now to the proof of the theorem and there replacing T, S_2, S_3, S_4 by t_{11}, T_2, T_3, T_4, respectively, we see that $T^T T \equiv I \pmod 2$ implies that $T_1^T T_1 \equiv I \pmod 2$. If p is odd, this process may be continued to prove the lemma. For $p = 2$ other cases need to be considered. In what follows we assume $p = 2$.

Secondly, suppose that t_{ii} is a unit for every i and $t_{ij}t_{ji}$ is a unit for some i and j, $i \neq j$. Permute rows and columns, if necessary, to make $t_{12}t_{21}$, a unit, and choose $D = I_2 \dotplus D_2$, where I_2 is the 2-rowed identity matrix, and let

$$\mu = \begin{bmatrix} t_{11} + 1 & t_{12} \\ t_{21} & t_{22} + 1 \end{bmatrix}.$$

Then we notice that μ is unimodular and set

$$P = \begin{bmatrix} I_2 & -\mu^{-1}T_2 \\ 0 & I \end{bmatrix} \quad \text{and} \quad T = \begin{bmatrix} T_0 & T_2 \\ T_3 & T_4 \end{bmatrix}$$

where T_0 is a 2 by 2 matrix, T_2 is 2 by $n - 2$, etc., Then

$$(T + D)P = \begin{bmatrix} T_0 + I_2 & 0 \\ T_3 & T_1 + D_2 \end{bmatrix}$$

where $T_1 = T_4 - T_3\mu^{-1}T_2$ and D_2 is to be chosen an automorph of $F = a_3 \dotplus a_4 \dotplus \cdots \dotplus a_n$. As in the preceding case we can replace T_3 by 0 and have $T_1^T T_1 \equiv I \pmod 2$.

Third, suppose T has the property that t_{ii} is a unit for every i, t_{ij} is a unit for some $i \neq j$ and, for every $i \neq j$, $t_{ij}t_{ji}$ is a non-unit. Now $T^T T \equiv I \pmod 2$ implies that each row and column of T contains an odd number of units and that for each i and j, $i \neq j$, there is an even number of values of k such that $t_{ik}t_{jk} \equiv 1 \pmod 2$. Thus, by a permutation of rows and columns we may assume that the leading 3 by 3 minor of T is congruent to

$$\begin{bmatrix} 1 & 1 & 1 \\ 0 & 1 & 1 \\ 0 & 0 & 1 \end{bmatrix} \pmod 2.$$

Furthermore, two of the first three diagonal elements of F are congruent (mod 4) and by a permutation of rows and columns of F and T we may assume that a_1 and a_2 are congruent (mod 4), that $t_{11}t_{12}t_{22}$ is a unit and t_{21} a non-unit. We can complete the proof along the lines of the previous case if we can find an automorph D_0 of $a_1x_1^2 + a_2x_2^2$ where

$$D_0 \equiv \begin{bmatrix} 0 & 1 \\ 1 & 0 \end{bmatrix} \pmod 2,$$

since then

$$\mu = D_0 + \begin{bmatrix} t_{11} & t_{12} \\ t_{21} & t_{22} \end{bmatrix}$$

will be unimodular. Suitable matrices D_0 are

$$\begin{bmatrix} 0 & \sigma^{-1} \\ \sigma & 0 \end{bmatrix} \quad \text{or} \quad \begin{bmatrix} 2 & 3\tau^{-1} \\ \tau & 2 \end{bmatrix}$$

according as $a_1 \equiv a_2 \pmod 8$ or $-3a_1 \equiv a_2 \pmod 8$, where σ and τ are p-adic units satisfying the equations $a_1 = a_2\sigma^2$, $-3a_1 = +a_2\tau^2$.

Finally suppose $T \equiv I \pmod 2$. Choose the ambiguous sign so that $t_{11} \pm 1$ is twice a unit and $D = \pm 1 \dotplus D_1$. By adding appropriate multiples of the first column of $T + D$ to the other columns and then similarly for rows we can reduce the first row and column of $T + D$ to $(t_{11} \pm 1, 0, 0, \cdots, 0)$. The remaining elements still will have the property that those on the diagonal are units and the non-diagonal elements are non-units.

By continuing the above reductive processes we can reduce $T + D$ to a direct sum of matrices $T_i + D_i$ where each $T_i + D_i$ is of one of three types: a unit, twice a unit, a 2 by 2 unimodular matrix. Hence $2(T + D)^I$ will be in $R(2)$.

Theorem 37 implies the following theorem for p odd. Further proof is necessary if p = 2.

THEOREM 38. *If the form f represents (is equivalent to) the form g in $R(p)$, $f = f_0 + f_1 + pf_2$, $g = f_0 + g_1 + pg_2$ where f_0, f_1, g_1 have unit determinants, g_1 and f_1 having the same number of variables, then $f_1 + pf_2$ represents (is equivalent to) $g_1 + pg_2$ in $R(p)$ provided that for $p = 2$, f_1 and g_1 are both properly primitive or both improperly primitive.*

First we may assume for $p = 2$, f_0, f_1, g_1 to be in the canonical forms of theorem 33a. Let r be the number of variables in f_1 and g_1. If $r = 0$, theorem 37 implies our result. Suppose f_1 and g_1 are the properly primitive forms

$$\sum_{i=1}^{r} \alpha_i x_i^2, \qquad \sum_{i=1}^{r} \beta_i y_i^2 .$$

If $r > 3$, theorem 17 shows that f_1 represents β_1 in $R(2)$ and hence by the proof of theorem 36 is equivalent to a form $\beta_1 y_1^2 + f_1'$ where f_1' represents a unit. This reduces consideration to $r \leqq 3$ since $f_1 + 2f_2 \cong \beta_1 y_1^2 + f_1' + 2f_2'$,

$g_1 + 2g_2 \cong \beta_1 y_1^2 + g_1' + 2g_2$ and the $\beta_1 y_1^2$ may be considered part of f_0, while if $f_1' + 2f_2'$ represents $g_1' + 2g_2$, $f_1 + 2f_2$ will represent $g_1 + 2g_2$. If $r = 2$, $f_1 \equiv \beta_1$ or $\beta_2 \pmod 8$ is solvable unless $\alpha_1 \equiv \alpha_2 \not\equiv \beta_1 \equiv \beta_2 \pmod 4$. In the former case $f_1 = \beta_1$ is solvable in $R(2)$ and we may proceed as above. In the latter case, suppose $2f_2 \equiv 2 \pmod 4$ solvable; then $f_1 + 2f_2 \equiv \beta_1$ or $\beta_2 \pmod 8$ is solvable and we reduce to the former case; similarly if $2g_2 \equiv 2 \pmod 4$ solvable. It remains to consider $2g_2 \equiv 2f_2 \equiv 0 \pmod 4$. Then $f \cong g$ and an equation similar to (17) shows that $f_0 + f_1 \cong f_0' + f_1' \equiv f_0 + g_1 \pmod 4$ which implies $| f_1 | \equiv | g_1 | \pmod 4$ and $c_2(f_0 + f_1) = c_2(f_0' + f_1') = c_2(f_0 + g_1)$ which implies $c_2(f_1) = c_2(g_1)$; these deny $\alpha_1 \equiv \alpha_2 \not\equiv \beta_1 \equiv \beta_2 \pmod 4$. If $r = 3$ or $r = 1$ we similarly may dispose of all cases.

Second if $p = 2$, and f_1 and g_1 are improperly primitive, take f_1 and g_1 in canonical form and see that either $2f_2 \equiv 2g_2 \equiv 0 \pmod 4$ in which case $f_1 \equiv g_1 \pmod 4$ and corollary 36c applies or $2f_2 \equiv 2 \pmod 4$ is solvable and f_2 may be utilized as above to make $| f_1 | \equiv | g_1 | \pmod 4$ and corollary 36c applicable.

As a matter of fact, in proving this theorem we have laid the basis for the proof of

THEOREM 38a. *If f_1, f_2, g_1, g_2 are forms of unit determinant in n_1, n_2, m_1, m_2 variables respectively, then $f = f_1 + 2f_2 \cong g = g_1 + 2g_2$ if and only if the following conditions hold.*

1. *$f \sim g$ and $n_i = m_i$, $i = 1, 2$.*
2. *f_i and g_i both properly or both improperly primitive, $i = 1, 2$.*
3. *If g_2 and f_2 are improperly primitive, $c_2(f_1) = c_2(g_1)$.*

The necessity follows from theorem 36. To see the sufficiency notice that the conditions above and the proof of theorem 38 imply the existence of a form $f_1' + 2f_2' \cong$

$f_1 + 2f_2$ with $f_1' \cong g_1$. Then $f_1' + 2f_2' \sim g_1 + 2g_2$ implies $f_2' \sim g_2$ and by theorem 36 $f_2' \cong g_2$ which shows $f_1' + 2f_2' \cong g_1 + 2g_2$ and $f \cong g$.

For an odd prime, the matter of equivalence is completely taken care of by the following theorem.

THEOREM 39. *If p is an odd prime and f and g are two forms in $R(p)$ written in the form of theorem* 35, *then $f \cong g$ if and only if $f_i \sim g_i$ in $R(p)$ for all i, that is, if and only if $|f_i| \sim |g_i|$.*

The sufficiency condition follows since $f_i \sim g_i$ implies, by theorem 36 that $f_i \cong g_i$. Then suppose $f \cong g$. Equation (17) with 2 replaced by p shows that a unimodular transformation takes f_1 into a form $f_1' \equiv g_1 \pmod{p}$ and hence by corollary 36a, $f_1' \cong g_1$ which implies $f_1 \cong g_1$. Theorem 37 then applies to show that $f - f_1 \cong g - g_1$ and the same process may be applied to prove $f_2 \cong g_2$. A continuation of this process proves the theorem.

Though the criteria for equivalence in $R(2)$ are not as simple as for $R(p)$ if p is odd, the above results can be used rather expeditiously in testing for equivalence as we show by the following example:

Let $f = f_1 + 2f_2 + 4f_3$, $\quad g = g_1 + 2g_2 + 4g_3$, where

$$f_1 = 6x_1^2 + 2x_1x_2 + 2x_2^2 + 6x_3x_4,$$

$$g_1 = 14y_1y_2 + 10y_3y_4,$$

$$f_2 = 17x_5^2 + 2x_5x_6 + 6x_6^2, \; g_2 = y_5^2 + 5y_6^2,$$

$$f_3 = 2x_7^2 + 6x_7x_8 + 6x_8^2 + 10x_9x_{10},$$

$$g_3 = 2y_7^2 + 6y_7y_8 + 2y_8^2 + 10y_9^2 + 6y_9y_{10} + 14y_{10}^2$$

First we show that $f \sim g$. From corollary 36a any unit coefficient of f or g may be replaced by one congruent to it (mod 8). Hence $f_2 \cong x_5^2 + 2x_5x_6 + 6x_6^2 = (x_5 + x_6)^2 +$

$5x_6^2 \cong g_2$. The determinants of $f_1 + f_3$ and $g_1 + g_3$ are congruent (mod 8) showing that $f_1 + f_3 \sim g_1 + g_3$ and $f \sim g$.

Now $|f_1| \not\cong |g_1|$ and hence $f_1 \not\cong g_1$. But we shall show that $f_1 + 2f_2 \cong f_1' + 2f_2'$ where $|f_1'| \equiv |g_1|$ (mod 8) and hence $f_1' \cong g_1$. If we let $f_0 = 6x_1^2 + 2x_1x_2 + 2x_2^2 + 2x_5^2$ and $g_0 = 14y_1y_2 + 10y_6^2$ we see that $c_2(f_0) = 1 = c_2(g_0)$ and $|f_0| = 22 \equiv |g_0|$ (mod 16) showing that $f_0 \sim g_0$ and hence, by theorem 38a, $f_0 \cong g_0$. Hence $f \cong g_0 + 6x_3x_4 + 10x_6^2 \cong f_1' + 2f_2' + 4f_3$ where $f_1' = 14x_1x_2 + 6x_3x_4 \cong g_1$ and $f_2' = 5x_5^2 + 5x_6^2$. Hence $f_2' + 2f_3 \sim g_2 + 2g_3$ from corollary 8 and, from theorem 38a, $f_2' + 2f_3 \cong g_2 + 2g_3$. But $f_1' \cong g_1$.

Thus we have shown that $f \cong g$ in $R(2)$.

28. **Representation of one form by another and the existence of forms with given invariants.** Some of the theorems in the previous section give partial criteria for the representation of one form by another. But it would seem that complete criteria are too complex to be included here. It is not true even for odd primes p that if f represents a form g there is a form h such that $f \cong g + h$ in $R(p)$. For instance $f = x_1^2 + 2x_2^2$ represents $3y_1^2$ in $R(3)$ but it is not equivalent to a form $3y_1^2 + ay_2^2$ for any a in $R(3)$. However if p is prime to $2|f||g|$, corollary 17 shows that f represents g in $F(p)$ if the latter has fewer variables than the former. Hence f represents the leading coefficient of g in $F(p)$ and therefore in $R(p)$. Thus $f \cong ax_1^2 + f'$ and $g \cong ay_1^2 + g'$ in $R(p)$. Then, by theorem 37, f' represents g' in $R(p)$ and so we continue. Hence *in corollary* 17, *$F(p)$ can be replaced by $R(p)$ if p is prime to* $2|f||g|$.

However, criteria for the existence of forms of given invariants are not hard to find. If p is odd, suppose the invariants t_i, f_i and m_i are given as in theorem 35, where m_i is the number of variables in f_i. We then merely choose

$$f_i = x_1^2 + x_2^2 + \cdots + x_{m_i-1}^2 + |f_i| x_{m_i}^2 .$$

Suppose $p = 2$ and we have a set of quantities t_i, m_i, $|f_i|$, $c_2(f_i)$ as in theorem 35 and specify that certain f_i shall be properly primitive and certain f_i improperly primitive. Notice that theorem 35 shows that t_i, m_i and the properly or improperly primitive conditions are invariants under unimodular transformations in $R(2)$; the other two quantities are not invariants. The existence of a form having the given values for the above quantities depends (see section 13) on the satisfaction of the following relationships:

i) For every i for which $m_i = 1$ or 2, the conditions of theorem 16 must be satisfied.

ii) If f_i is an improperly primitive form of determinant d_i in m_i variables, then $m_i = 2w_i$ and either $d_i = (-1)^{w_i}$ with $c_2(f_i) = k_i$ or $d_i = 3(-1)^{w_i-1}$ with $c_2(f_i) = -k_i$, where

$$k_i = (-1)^{(w_i-2)(w_i-1)/2}.$$

If all relationships i) and ii) hold, the form exists in $R(2)$. (See section 34.)

29. **Zero forms and universal forms.** If f is a form in $R(p)$ and if $f = 0$ has a non-trivial solution in $F(p)$ then, by multiplying this solution by a power of p, we get a non-trivial solution of $f = 0$ in $R(p)$. Thus the conditions that f be a zero form or, in fact, that it be an m-zero form, are exactly the same as for $F(p)$.

As to the universality of zero forms, consider first p odd and f written in the form $f = f_1 + p^{t_2}f_2 + \cdots + p_k^{t_k}f_k$ where each f_i has unit determinant. Let $(r_1, r_2, \cdots, r_n)$ be a *primitive solution* of $f = 0$, that is one in which 1 is the g.c.d. of the r_i. If the first m_1 numbers r_i are divisible by p, then $f' = 0$ has a non-trivial solution $(r_1/p, r_2/p, \cdots, r_{m_1}/p, r_{m_1+1}, \cdots, r_n)$ where $f' = pf_1 + p^{t_2-1}f_2 + \cdots + p^{t_k-1}f_k$. So we proceed until we have a form $f_0 = f_{10} + p^{s_2}f_{20} + \cdots + p^{s_k}f_{k0}$ such that $f_0 = 0$ has a solution $(u_1, \cdots, u_n)$

with not all of $u_1, \cdots, u_n$ divisible by p, n_1 being the number of variables in f_{10}. Then the proof of theorem 13 shows that f_0 is universal. We cannot, however, be sure that f is universal unless $f_1 \equiv 0 \pmod p$ has a non-trivial solution. For example $f = x_1^2 + x_2^2 + 9x_3^2 + 9x_4^2$ is a zero form in $R(3)$ but it is not universal since it does not represent 6 in $R(3)$; but $f' = x_1^2 + x_2^2 + x_3^2 + x_4^2$ is universal in $R(3)$. On the other hand, $f = x_1^2 + x_2^2 + 3x_3^2 - 3x_4^2$ is universal even though $f = 0$ has no solution with x_1 or x_2 prime to 3. Along these lines one could formulate general criteria.

For $p = 2$, the same process as that above may be carried through until we have f_0 having the above properties. Then the proof of theorem 13 tells us that $f_0 = 4N$ is solvable for all N and the same process may be continued to get an f_0' representing all N.

Similar conditions hold for m-universality.

30. **Automorphs and representations.** If X_1 and X_2 are two unimodular transformations taking A into B we can see exactly as in the proof of theorem 7 that there is an automorph U of A such that $UX_1 = X_2$. If p is odd, X_1 and X_2 are two n by m matrices, $n > m$, such that $X_1^TAX_1 = X_2^TAX_2 = B$, if X_1 and X_2 each have an m-rowed unit minor determinant and if B has unit determinant, the proof of theorem 7 carries through to prove a similar result since theorem 37 implies the result corresponding to corollary 8. That both conditions p odd and B of unit determinant are necessary is shown by the following examples.

Example 1. Notice that $v = (1, 0, 0)$ and $w = (1, -1, 0)$ are two solutions of $f = 1$ in $R(2)$ where $f = x_1^2 + 2x_2x_3$. If $U = (u_{ij})$ were an automorph of f such that $Uv^T = w^T$ then $1, -1, 0$ would be the first column of U. The fact that

U is an automorph implies that u_{12} and u_{13} are both even and $u_{22}u_{33} - u_{23}u_{32}$ is odd. Substitution shows that there is no such automorph.

EXAMPLE 2. Here $v = (1, 1, 0)$ and $w = (0, 0, 1)$ are solutions of $f = 3$ in $R(3)$ where $f = x_1^2 + 2x_2^2 + 3x_3^2$. Then $Uv^T = w^T$ implies $u_{11} + u_{12} = 0 = u_{21} + u_{22}$, $u_{31} + u_{32} = 1$. Then U an automorph implies $u_{12}^2 - u_{22}^2 \equiv 1 \pmod 3$ on the one hand and $u_{12}^2 - u_{22}^2 \equiv 2 \pmod 3$ on the other. Thus there is no such automorph.

31. **Binary Forms.** In this section we particularize our results on equivalence in $R(p)$ to the binary case. These results could be shown, of course, without the general theory but with much more labor. We may assume the form written: $f = ax^2 + 2bxy + cy^2$ where a, b, c are integers. Furthermore f may be taken to be primitive since its p-adic class is determined by g, the g.c.d. of a, b, c, and the class of f/g. Write $|f| = d$.

If p does not divide $2d$, corollary 36b shows that the class is determined by $(d \mid p)$.

If p is an odd prime dividing d, we may assume f to be of the form $ax^2 + p^k c_0 y^2$ by theorem 32 and, using theorem 39, d and $(a \mid p)$ determine the class of f.

If $p = 2$ and the form is improperly primitive, $d \equiv 3 \pmod 4$ and the 2-adic class is determined by $d \pmod 8$. If the form is properly primitive it may by theorem 33a be taken into a form $ax^2 + 2^k c_0 y^2$. If $k < 2$, use theorem 38a, if $k \geqq 2$ use theorem 9b or other considerations to show that d and the following determine the class:

1. If $d \equiv 3 \pmod 4$, no other condition.
2. If $d \equiv 1 \pmod 4): a \pmod 4$.
3. If $d \equiv 2 \pmod 8): (-2 \mid a)$.
4. If $d \equiv 6 \pmod 8): (2 \mid a)$.

5. If $d \equiv 4 \pmod{8} : a \pmod{4}$.
6. If $d \equiv 0 \pmod{8} : a \pmod{8}$.

Notice that the number of classes of given determinant in $R(p)$ is 1 if p is odd and prime to the determinant or $p = 2$ and $d \equiv 3 \pmod{4}$ with f properly primitive or f improperly primitive; 2 if p is odd and divides the determinant or $p = 2$ and $d \equiv 1, 2 \pmod{4}$ or $\equiv 4 \pmod{8}$; 4 if $p = 2$ and $d \equiv 0 \pmod{8}$. Furthermore in all cases the class in $R(p)$ is determined by any number prime to p represented by the form.

CHAPTER V

GENERA AND SEMI-EQUIVALENCE

32. **Definitions.** Reasoning by analogy from the results of chapter III one might deduce that the function of this chapter would be to prove that if a form f with integral coefficients represents a number N in $R(p)$ for all p and in the field of reals, then there would be integer values of the variables of f for which $f = N$. One might also suppose that a similar result would hold for equivalence in $R(p)$ and in the ring of integers. But, while it is true that equivalence (or representation) in the ring of integers implies equivalence in $R(p)$ for all p, yet the converse statement is not true as is shown, for instance, by the fact that 8/5, 1/5 is a solution of $f = x^2 + 11y^2 = 3$ in the field of reals, in $R(2)$, $R(3)$ and $R(11)$. Thus f represents 3 in all $R(p)$, from corollary 14 and theorem 34, but $f = 3$ has no solution for integer values of x and y. However, two things do follow from the fact that f represents 3 in all $R(p)$. First, for any integer q, there is a solution of $f = 3$ in rational numbers with denominators prime to q. Second, there is a form g with integer coefficients such that $g = 3$ has an integral solution and such that for every integer q there is a transformation which takes f into g and whose elements are rational numbers with denominators prime to q. Here $g = 3x^2 + 2xy + 4y^2$ and, for instance, if q is prime to 5, the transformation

$$\begin{bmatrix} 8/5 & -1/5 \\ 1/5 & 3/5 \end{bmatrix}$$

takes f into g.

Using a term of Siegel's we say that an n-ary form f with matrix A represents an m-ary form g $(n \geqq m)$ of matrix B

rationally without essential denominator if, for every positive integer q, there is a matrix T such that $T^TAT = B$, T is of rank m and has rational elements with denominators prime to q. If 1 is the g.c.d. of the m-rowed minor determinants of T, we call the representation *primitive*. If $n = m$ and each form represents the other rationally without essential denominator we call the two forms *rationally equivalent without essential denominator* or, more briefly, *semi-equivalent* and say that f and g are in the same *genus*, writing $f \text{ v } g$. Two forms in the same genus must have equal determinants. In this chapter the terms *equivalent* and *represents* unadorned are understood to refer to equivalence and representation in the ring of integers. All forms are assumed to be "integral forms" that is, having matrices with integral elements, except where something is said to the contrary. We call a matrix *q-rational* if its elements are rational numbers with denominators prime to q. Similarly we speak of *q-rational representation*.

Using the above terminology we may consider this chapter devoted chiefly to proving that if f represents g in $R(p)$ for all p including $p = \infty$, then first, f represents g rationally without essential denominator and, second, there is a form f_0 in the genus of f which represents g in the ring of integers.

33. **Representation without essential denominator.** There are three ways in which semi-equivalence may be defined and each definition has its importance. (As a matter of fact the first definition of semi-equivalence was different from these three. But for forms of more than three variables it is very cumbersome and we here omit it.) Hence we devote this section chiefly to proving the logical equivalence of the statements embodied in

THEOREM 40. *If f and g are two integral forms with n and*

m variables respectively, any one of the following statements implies all the others where, if $m = n$, the word "represents" is to be replaced by "is equivalent to". Furthermore if in 2b *and* 3b *"represents" is replaced by "primitively represents" the two statements are logically equivalent.*

1a. *Form f represents g rationally without essential denominator.*

1b. *Form f represents g q-rationally with $q = 2|f||g|$.*

2a. *Form f represents g in $R(p)$ for all p including $p = \infty$.*

2b. *Form f represents g in $R(p)$ for all primes p dividing $2|f||g|$ and $p = \infty$.*

3a. *For q an arbitrary integer, f represents $g_0 \equiv g$* (mod q), *f represents g in $R(\infty)$ and, if $m = n$, $|f| = |g|$.*

3b. *Form f represents $g_0 \equiv g$* (mod $8|g|P$) *where P is the product of the distinct odd primes in $|f||g|$, f represents g in $R(\infty)$ and, if $m = n$, $|f| = |g|$.*

To prove the logical equivalence of the first four statements it is sufficient to show that 1a implies 1b implies 2b implies 2a implies 1a. Now 1b is a special case of 1a. Furthermore, 1b implies 2b since if f represents g p-rationally, the elements of the transformation are integers in $R(p)$ and hence f represents g in $R(p)$. If $m = n$, statement 2b implies $|f| = |g|$ in $R(p)$ for all primes p dividing $2|f||g|$ and for $p = \infty$; hence $|f| = |g|$ and, from theorem 36, statement 2a holds; while if $n > m$, section 28 has the same effect. The fact that 2a implies 1a is the important result contained in theorem 42 below.

The proof of our theorem will now be complete if we establish the following chain of implication: 2a implies 3a implies 3b implies 2b. By lopping off, after a finite number of terms, the p-adic expansion of any representation of g by f given by statement 2a, we see that if T_p is a representation in $R(p)$ of g by f, then for t arbitrary there is a matrix $T'_{p^t} \equiv T_p$ (mod p^t) and having integer elements.

In fact, by the Chinese remainder theorem, there is a matrix T'_q with integer elements which is congruent to $T_p \pmod{q}$ for every prime factor p of q. Furthermore if T_p is a primitive solution for all p dividing q or if $m = n$, lemma 9 below shows that T'_q may be chosen to have 1 as the g.c.d. of its m-rowed minor determinants. Then T'_q takes f into a form $g_0 \equiv g \pmod{q}$. Since 2a implies that $|f| = |g|$ if $m = n$ we have completed the proof that 2a implies 3a. Furthermore a primitive representation in 2b implies that in 3b by the above proof. Statement 3b is a special case of 3a.

It remains to prove that statement 3b implies 2b. To this end let p be a prime dividing $2|f||g|$. Then statement 3b implies that there is an integral matrix T taking f into $g_0 \equiv g \pmod{p^{u+w}}$ where p^u is the highest power of p in $|g|$ and w is 1 or 3 according as p is odd or $p = 2$. Hence theorem 9a shows that f represents g in $R(p)$.

In the above proof we used the following lemma.

LEMMA 9. *If q is a positive integer, if T is an n by m matrix, $n \geqq m$, with integer elements and if the* g.c.d. *of the m-rowed minor determinants of T is congruent to $g \pmod{q}$ where g is an integer prime to q, then there is a matrix S with integer elements congruent to $T \pmod{q}$ such that g is the* g.c.d. *of its m-rowed minor determinants. If $m = n$, the* g.c.d. *is understood to be the value of the determinant.*

To prove this first notice that by lemma 4, there are unimodular matrices P and Q such that $PTQ = R = (r_{ij})$ has $r_{ii} = r_i$, $r_{ij} = 0$ for $i \neq j$ and, by lemma 5, the product of the r_i congruent to $g \pmod{q}$. Now r_{m-1} may be chosen $\pmod{q}$ prime to r_m; r_{m-2} chosen prime to r_m and r_{m-1}, $\cdots$, and r_1 prime to $r_2, r_3, \cdots, r_m$ and so that

$$h = \prod_{i=1}^{m} r_i = g + q^2 L$$

define integers h and L; this may be done since the product of the r_i is prime to q. Thus we assume that the r_i are relatively prime in pairs and their product equal to $g + q^2L$. Then in R replace all but the first element in the first row by q and make the first m elements of the first column: $r_1, k_2q, \cdots, k_mq$, the k_i integers later to be chosen. This yields a matrix $R' \equiv R \pmod{q}$, the determinant of whose first m rows is

$$h - q^2 \sum_{i=2}^{m} k_i\, h/r_i\, r_1 = g + q^2 \left(L - \sum_{i=2}^{m} k_i\, h/r_1 r_i\right).$$

Now for any s and j, $s \neq j$, a common factor of $h/r_i r_j$ and $h/r_i r_s$ is a factor of $h/r_i r_j r_s$. Hence any common factor of the product of $m - 2$ of $r_2, r_3, \cdots, r_m$ is a common factor of all products of $m - 3$ of the same set of r_i. Continuing in this fashion we see that, since the r_i are prime in pairs, the g.c.d. of the coefficients of k_i in the sum above have 1 as their g.c.d. Hence the k_i may be chosen so that the sum is L. Then the determinant of the first m rows of R' is g and thus, by lemma 5, the g.c.d. of the m-rowed minor determinants of $S = P^I R' Q^I$ is g, while $R' \equiv R \pmod{q}$ implies $S \equiv T \pmod{q}$.

This lemma is also of use in proving a result corresponding for $p = 2$ to that of theorem 25 for p odd. From theorem 33 there is a transformation unimodular in $R(2)$ taking any form f into a form

$$f_0 = 2^{t_1}g_1 + 2^{t_2}g_2 + \cdots + 2^{t_r}g_r$$

where each g_i is a form with unit determinant in $R(2)$ and hence, by theorem 33a is of one of the types there described. Suppose f has integer coefficients and T is a transformation unimodular in $R(2)$ taking f into f_0. Let $|T|^{-1} = u$, u being a unit in $R(2)$. If we multiply the first column of T by u, the first variable in g_1 is multiplied by the unit u. After this change g_1 will still have unit determinant and be of the type described in theorem 33a. Thus we may assume that the transformation T in $R(2)$ taking f into f_0 has determinant 1. Then for any positive integer k we may consider each 2-adic element of T expanded into a series and, deleting the portion of each series after the term containing 2^k we have a matrix T_k with integral elements and congruent to T (mod 2^k). Then, by lemma 9, there is a unimodular transformation T_0 congruent to T_k and T (mod 2^k). This will take f into an integral form f' congruent to f_0 (mod 2^k) and proves the theorem below for $p = 2$. For p an odd prime a similar proof may be used or theorem 25. We have then

THEOREM 41. *Any form f with integral coefficients is, for an arbitrary positive integer k and prime p, congruent* (mod p^k) *to an integral form*

$$f = p^{t_1}g_1 + p^{t_2}g_2 + \cdots + p^{t_r}g_r$$

where each g_i is an integral form with determinant prime to p and diagonal unless $p = 2$ in which case it may be assumed to be one of the types described in theorem **33a.**

We must prove

THEOREM 42. *If an n-ary integral form f represents an m-ary integral form g ($n \geqq m$) in $R(p)$ for all p including $p = \infty$, then f represents g rationally without essential denominator. In particular if $n = m$ and "represents" is replaced by "is equivalent to" we can conclude that f v g.*

We need the following two lemmas for this theorem.

LEMMA 10. *Let F be any field, A and B two symmetric matrices in F with n and m rows respectively, $n \geqq m$, and T_0 an n by m matrix in F such that $T_0^T A T_0 = B$. Then for any matrix T in F satisfying the equation $T^T A T = B$, and for which $(T_0^T A T - B)^I$ exists, there is a skew symmetric* *matrix Q in F and a matrix P such that*

$$(22) \quad T = MT_0, \qquad M = I + 2P(Q - P^TAP)^IP^TA.$$

If, conversely, Q is any skew symmetric matrix in F and P an arbitrary matrix in F for which $(Q - P^TAP)^I$ exists, then (22) defines a matrix T satisfying the equation $T^TAT = B$. The matrix M is an automorph of A.

Let X be defined by the equation $T = T_0 + X$. Then $T^TAT = B$ becomes

$$X^TAT_0 + T_0^TAX = -X^TAX.$$

Replacing X by PY^I, multiplying on the right by Y and on the left by Y^T converts this equation into

$$P^TAT_0Y + Y^TT_0^TAP = -P^TAP.$$

Hence the matrix

$$(23) \qquad Q = 2P^TAT_0Y + P^TAP$$

is skew symmetric. Since $(T_0^TAT - B)^I$ exists by hypothesis and $T_0^TAT - B = T_0^TAX = T_0^TAPY^I$, we see tht P^TAT_0Y is non-singular and hence $Q - P^TAP$ is also. Then notice that

$$(Q - P^TAP)^I2P^TAT_0Y = I,$$

multiply on the right by Y^I and on the left by P and see that $X = PY^I$ implies (22).

Conversely if $(Q - P^TAP)^I$ exists, multiplication shows that M is an automorph of A and hence that (22) defines a matrix T for which $T^TAT = B$. (This lemma and proof are Siegel's).[10]

LEMMA 11. *Let R be an n by n matrix in $F(p)$ for some prime p, B a non-singular n by n symmetric matrix in $R(p)$. There is an automorph D of B in $R(p)$ such that* $|RD - B| \neq 0$. (Cf. lemma 8)

To prove this notice from theorems 33 and 33a that B is equivalent in $R(p)$ to a diagonal form if p is odd or, when $p = 2$, to a direct sum of matrices of the three following types:

$$a, \begin{bmatrix} 2 & 1 \\ 1 & 2 \end{bmatrix}, \quad \begin{bmatrix} 2 & 5 \\ 5 & 12 \end{bmatrix}$$

since

$$\begin{bmatrix} 1 & 2 \\ 1 & 3 \end{bmatrix} \quad \text{takes} \quad \begin{bmatrix} 0 & 1 \\ 1 & 0 \end{bmatrix} \quad \text{into} \quad \begin{bmatrix} 2 & 5 \\ 5 & 12 \end{bmatrix}.$$

These three types have the respective automorphs:

$$\pm 1; \pm \begin{bmatrix} 1 & 0 \\ 0 & 1 \end{bmatrix}, \pm \begin{bmatrix} 1 & 1 \\ 0 & -1 \end{bmatrix}; \pm \begin{bmatrix} 1 & 0 \\ 0 & 1 \end{bmatrix}, \pm \begin{bmatrix} 1 & 5 \\ 0 & -1 \end{bmatrix}.$$

Let D_i, $i = 1, \cdots, s$ be the matrices obtained from B by replacing each component of B by one of its automorphs listed above. We see that the D_i form a *group* under multiplication that is, satisfy for multiplication properties 1, 3, 4, 7 listed in section 7, and that $s = 2^n$. Let K be that diagonal matrix whose elements are the indeterminates $\lambda_1, \lambda_2, \cdots, \lambda_n$ and consider the sum

$$L = \sum_{i=1}^{s} | RKD_i - B |.$$

This is a linear function of each of the λ_i and is left unaltered if each D_i is replaced by $-D_i$. Thus L is an even function of each λ_i and hence is independent of λ_i. Thus taking $K = I$ and $K = 0$, we have

$$L = \sum_{i=1}^{s} | RD_i - B | = s\,|-B| \neq 0$$

which shows that one of the determinants in the sum is non-singular. (The methods of this proof are extensions of those Siegel[14] used to prove a similar result.)

Now we proceed to prove theorem 42. The hypothesis of our theorem and theorem 31 shows us that there is a transformation T_0 with rational elements taking A into B where A and B are the matrices of f and g respectively. There is also a matrix T_p in $R(p)$ taking A into B. Taking R in lemma 11 to be $T_0^T A T_p$ we see that, by replacing T_p by $T_p D$ for some automorph D of B we may assume that $| T_0^T A T_p - B | \neq 0$. This replacement does not alter the fact that $T_p^T A T_p = B$.

Let q be any given positive rational integer and p a prime factor of q. Then lemma 10 shows that there is a skew-symmetric matrix Q_p in $F(p)$ and a matrix P_p in $F(p)$ such that

$$T_p = T_0 + 2P_p(Q_p - P_p^T A P_p)^I P_p^T A T_0 .$$

If b is an arbitrarily large rational integer, by the Chinese remainder theorem we can find matrices P and Q, the latter being skew-symmetric and both having rational elements, satisfying the congruences $P \equiv P_p \pmod{p^b}$, $Q \equiv Q_p \pmod{p^b}$ for all prime factors p of q. Now $| T_0^T A T_p - B | \neq 0$ implies $| Q_p - P_p^T A P_p | \neq 0$. Hence, for b sufficiently large $| Q - P^T A P | \equiv | Q_p - P_p^T A P_p | \not\equiv 0 \pmod{p^b}$ for all p dividing q. Hence $| Q - P^T A P | \neq 0$. Then define the matrix T with rational elements by the following equation

$$T = T_0 + 2P(Q - P^T A P)^I P^T A T_0 .$$

Since T_p has p-adic integers as elements and $T \equiv T_p \pmod{p^b}$ for b arbitrarily large, T has p-adic integers as elements for all primes p dividing q and hence the denominators of its elements are prime to q. This completes the proof of theorems 40 and 42.

The following theorem has important consequences.

THEOREM 43. *Let A and B be symmetric integral non-singular matrices with respective dimensions n and m ($n \geqq m$)*

and S an n by m matrix of rank m with rational elements such that s is the l.c.m. *of the denominators and $S^T AS = B$. Then there is an n by n matrix T with rational elements the prime factors of whose denominators all divide s, whose determinant is* 1 *and which takes A into an integral matrix A_0 which represents B integrally, that is, $U^T A_0 U = B$ for some integral matrix U.*

To prove this we first, for brevity's sake, define an *s-matrix* or *s-transformation* to be one with rational elements the prime factors of whose denominators all divide s. Then write $R = sS$ and, by lemma 4 determine unimodular matrices P and Q such that

$$PRQ = \begin{bmatrix} R_1 \\ 0 \end{bmatrix} = s \begin{bmatrix} S_1 \\ 0 \end{bmatrix} = sS'$$

where R_1 is the diagonal matrix $r_1 \dotplus r_2 \dotplus \cdots \dotplus r_m$, r_i divides r_{i+1} for $i = 1, 2, \cdots, m - 1$ and S' and S_1 are defined by the equations. Write $r_i/s = u_i/s_i$ where the latter fraction is in lowest terms and $s_i > 0$. Then s_i is divisible by s_{i+1} and hence s_i is prime to u_j for $j \leqq i$.

If we write $A' = P^{IT} AP^I$ and $B' = Q^T BQ$, the equality $S^T AS = B$ becomes $S'^T A'S' = B'$ which, with $A' = (a_{ij})$ and $B' = (b_{ij})$ implies

(24) $\quad a_{ij} r_i r_j / s^2 = a_{ij} u_i u_j / s_i s_j = b_{ij}, \; i, j = 1, 2, \cdots, m.$

Suppose $s_i = 1 = | u_i |$ for $i = 1, 2, \cdots, h - 1$ but not for $i = h$. Then (24) and B' integral implies that s_h divides a_{ih} for $1 \leqq i \leqq h$ and s_h^2 divides a_{hh}. Moreover u_h divides b_{ih} for $1 \leqq i \leqq m$ and u_h^2 divides b_{hh}. Write A' in the form

$$\begin{bmatrix} A_{11} & A_{12} \\ A_{12}^T & A_{22} \end{bmatrix}$$

where A_{11} is an h by h matrix. Let D_h be the matrix obtained from the h-rowed identity matrix by replacing its last diagonal element by s_h and U_h the matrix obtained from the m-rowed identity matrix by replacing its h-th diagonal element by u_h. Write $F_h = D^I \dotplus K_h$ where K_h is an integral $n - h$ rowed square matrix of determinant s_h later to be determined. We then have

$$F_h^T A' F_h = \begin{bmatrix} D_h^I A_{11} D_h^I & D_h^I A_{12} K_h \\ K_t^T A_{12}^T D_h^I & K_h^T A_{22} K_h \end{bmatrix} = A_h .$$

Then S_h, defined by the equation $S_h = F_h^I S' U_h^I$, is the direct sum of the integral matrix $u_1 \dotplus u_2 \dotplus \cdots \dotplus u_{h-1} \dotplus 1$, whose elements have absolute value 1 and an $n - h$ by $m - h$ s-matrix. Furthermore $S'^T A' S' = B'$ becomes $S_h^T A_h S_h = B_h$ where the equality $U_h^I B' U_h^I = B_h$ defines B_h. Now B_h is integral since, as shown above, u_h divides b_{ih} for $1 \leqq i \leqq m$ and u_h^2 divides b_{hh}. Furthermore B_h represents B' integrally. Similarly $D_h^I A_{11} D_h^I$ is integral.

Next we determine K_h consistent with the above conditions so that A_h is integral, that is, so that $D_h^I A_{12} K_h$ is integral. In fact, in view of the definition of D_h, we need only make αK_h divisible by s_h where α is the h-th row of A_{12}. This is easily done by finding a unimodular matrix W so that $\alpha W \equiv (w, 0, \cdots, 0) \pmod{s_h}$ and choosing $W^I K_h$ to be the diagonal matrix $s_h \dotplus 1 \dotplus \cdots \dotplus 1$ of determinant s_h.

If now we diagonalize the last $n - h$ rows of S_h replacing A_h and B_h by equivalent matrices we may continue along the above lines inductively to derive a sequence of integral matrices A_h obtained from A by s-transformations of determinant 1 and taken by transformations S_h having the properties described above into integral matrices B_h which, in turn, represent B' integrally. Then take $A_m = A_0$, $B_m = B_0$ and see that A_0 is obtained from A by an s-transformation of determinant 1, represents B_0 integrally and hence B integrally.

This completes the proof of theorem 43, whose chief

importance is in its application. Suppose now f represents g rationally without essential denominator. That means that there is a matrix with rational elements, the l.c.m. of whose denominators is s, an integer prime to $2 \mid f \parallel g \mid$ taking f into g. Then, by theorem 43 there is a transformation T whose elements are rational numbers the prime factors of whose denominators divide s, whose determinant is 1 and which takes f into an integral form f_0 which represents g integrally. Hence by theorem 40 and the definition of semi-equivalence we have

THEOREM 44. *If an integral form f represents an integral form (or integer) g in fewer variables rationally without essential denominator, there is an integral form f_0 which is semi-equivalent to f and which represents g integrally. Also if f represents g, there is an integral form f_0, with $f_0 \cong f$, which represents g integrally.*

The fact that statement 3b in theorem 40 implies 1a and the truth of theorem 44 yields

COROLLARY 44a. *If f represents integrally a form $g \equiv g_0$* (mod q) *where* $q \equiv 0$ (mod $8p \mid g \mid$) *and if f represents g_0 in the field of reals, then there is an f_0 semi-equivalent to f and representing g_0 integrally. P is the product of the odd prime factors of $\mid f \parallel g \mid$.*

Similarly we have

COROLLARY 44b. *If f represents a form g in $R(p)$ for all primes p dividing $2 \mid f \parallel g \mid$ and $p = \infty$ there is a form f_0 semi-equivalent to f and representing g integrally.*

Now let us apply theorem 44 to a particular example. Suppose $f = x_1^2 + x_2^2 + 3x_3^2$. First we show that every form in the genus of f is equivalent to it. This may be done most expeditiously by reference to a table[3,5] of reduced positive ternary quadratic forms but, since it is not very tedious to prove this fact directly for this simple form, we do it without use of such a table. Theorem 23 tells us that any positive integral form f of determinant 3 is equivalent to a form whose leading coefficient is $\leqq (4/3)\sqrt[3]{3}$, that is < 2. Hence by the same theorem, $f_0 = x_1^2 + f_1$ where f_1 is a positive binary form of determinant 3. By the same theorem f_1 represents 1 or 2; in the former case we have our given form, in the latter case $h = x_1^2 + 2x_2^2 + 2x_3^2 + 2x_2x_3$. Now $f \equiv 6 \pmod 9$ is not solvable whereas 2, 1, 0 is a solution of $h = 6$. Thus h is equivalent to a form h_0 whose leading coefficient is 6 and $f \equiv h_0 \pmod 9$ is not solvable. This shows from theorem 40 that f is not in the genus of h.

Since we have just shown that any form semi-equivalent to f is equivalent to f, corollary 44b shows that f represents a number a if and only if f represents a in the field of reals, that is, if and only if a is non-negative and f represents a in $R(p)$ for all primes p dividing 6a. Corollary 34b shows that only $p = 2$ and $p = 3$ need be considered. If $p = 2$, theorem 34a shows that we need merely show that $f = a$ is solvable in $F(2)$; but by section 13 $c_2(f) = 1$ and hence, by corollary 14, f represents a in $F(2)$. Furthermore $c_3(f) = -1$ and corollary 14 shows that f represents a in $F(3)$ if and only if $-3a$ is a non-square in $F(3)$, that is a is not of the form $9^k(9N + 6)$. But $f = a$ solvable in $F(3)$ implies that for some integer k, $f = 9^k a$ is solvable in $R(3)$. However $f = 9^k a$ implies that $x_1 \equiv x_2 \equiv x_3 \equiv 0 \pmod{3^k}$ and f represents a; that is, $f = a$ is solvable in $F(3)$ if and only if it is solvable in $R(3)$. Thus we have shown that f represents all positive integers except those of the form

$$9^k(9N + 6)$$

and f represents none of this form.

Notice that if, in corollary 44a and 44b, all forms in the genus of f are equivalent, the statements made about f_0 hold also for f.

34. Existence of forms with integral coefficients and having given invariants. In proving the main theorem of this section we shall need a lemma and an auxiliary theorem. First we prove

LEMMA 12. *If f is an n-ary form with integer coefficients and g a form in $R(p)$ such that $f \sim g$ in $F(p)$ and $|f|/|g|$ is a unit in $F(p)$, there is an integer K such that for every $k \geqq K$ there is a transformation T whose determinant is 1, whose elements are integers divided by powers of p and which takes f into an integral form $f_0 \equiv g'$ (mod p^k) where g' is obtained from g by replacing one of the variables x of g by ux, u being an integer prime to p.*

Let S be the transformation in $F(p)$ which takes f into g. Let K be an integer greater than the highest power of p in the denominators of the elements of S and greater than the highest power of p in $|f|$ and in $|g|$. Take $k \geqq K$ and deleting all terms after p^k in the expansion of the elements of S, we have a transformation $S' \equiv S$ (mod p^k) which takes f into a form $g_1 \equiv g$ (mod p^k). Since g is in $R(p)$ and all denominators of S' divide p^{k-1}, g_1 has integral coefficients. Then, applying theorem 43 we can obtain a transformation T' whose elements are integers divided by powers of p, whose determinant is 1 and which takes f into an integral form f' which represents g_1 integrally. Let U be the transformation taking f' into g_1. Now $|f| = |f'|$, $g_1 \equiv g$ (mod p^k) and $|f|/|g|$ a unit implies $|U|^{-1}$ is congruent to u (mod p^k), u being an integer prime to p. Then replacing the leading variable x by ux in g_1 and g to get g_1' and g' respectively, we see that f' is taken into g_1' by a transformation U' obtained from U by multiplying the first column by u. Now $|U'| \equiv 1$ (mod p^k) and hence, by lemma 9, there is an integral matrix U'' of determinant 1 and congruent to U' (mod p^k). Then set $T = T'U''$ and see that it not only takes f into a form $f_0 \equiv g'$ (mod p^k) but has determinant 1 and elements which are integers divided by powers of p. This completes the proof.

Next we need the following extension of theorem 29.

THEOREM 45. *If, in addition to the conditions of theorem 29, n is even and $d \equiv (-1)^{n/2} \pmod 4$, there is a properly primitive form and an improperly primitive form in n variables, having determinant d, index i and given Hasse invariants $c_p(f)$.*

The conditions of theorem 29 assure us of the existence of a form f having the invariants of that theorem. Then use theorem 41 to see that f is equivalent to one of the following:

$$f_1 \equiv a_1x_1^2 + a_2x_2^2 + \cdots + a_nx_n^2 \pmod 4, \qquad a_i = 1 \text{ or } -1$$

$$f_2 = \sum_{i=1}^{n/2} a_{2i-1}\, x_{2i-1}^2 + 2x_{2i-1}\, x_{2i} + a_{2i-1}\, x_{2i}^2 \pmod 4$$

where a_{2i-1} is 2 or 0.

First suppose $f \cong f_2$. Then

$$Q = \begin{bmatrix} 1 & -1 \\ 1 & 1 \end{bmatrix}$$

takes $2xy$ into $2(x^2 - y^2)$ and $2x^2 + 2xy + 2y^2$ into $6x^2 + 2y^2$. Thus the direct sum of $n/2$ transformations Q takes f_2 into a form $2f_3$ where f_3 is properly primitive. Since this transformation is in $R(p)$ for all odd p it leaves unaltered $c_p(f)$ for all p except 2 and hence $c_2(f) = c_2(f_3)$ shows that f_3 has the required invariants.

Second suppose $f \cong f_1$. In $n \geq 6$, four of the coefficients of f_1 must be congruent (mod 4). Then the transformation

$$\begin{bmatrix} 1 & 1 & 1 & 1 \\ 1 & -1 & 1 & 1 \\ 1 & 1 & -1 & 1 \\ 1 & 1 & 1 & -1 \end{bmatrix}$$

of determinant -8 takes $f_0 = x_1^2 + x_2^2 + x_3^2 + x_4^2$ into $2g$ where g is improperly primitive and has the same invariants as f_0. Then by a permutation of variables if necessary we may take f_i into a form congruent to $f_{01} + f_{02} + \cdots + f_{0k}$ (mod 4) where all but at most the last f_{0i} is congruent to $\pm(x_1^2 + x_2^2 + x_3^2 + x_4^2)$ (mod 4) and hence by the above transformation may be taken into a form $2g_{0i}$ where g_{0i} is improperly primitive. If f_{0k} has four variables, $| f | \equiv (-1)^{n/2}$ (mod 4) implies $f_{0k} \equiv \pm f_0$ (mod 4) or $f_{0k} = x_1^2 - x_2^2 + x_3^2 - x_4^2$ which is taken by the transformation $2(Q^I \dotplus Q^I)$ into twice an improperly primitive form. If f_{0k} has two variables $| f_{0k} | \equiv (-1)^{n/2}$ (mod 4) implies $f_{0k} \equiv x_1^2 - x_2^2$ (mod 4) and $2Q^I$ applies. Thus in all cases we may replace f_1 by $2f_4$ where f_4 has the same invariants as f. This completes the proof of the theorem.

Now theorems 28 and 15 show that two forms f and g of equal determinant are rationally congruent if the indices and number of variables of f and g are equal and $c_p(f) = c_p(g)$ for all p dividing $2\,|f|$. Hence we call $|f|$ except for a square unit factor, $c_p(f)$, i and n, the number of variables of f, its *field invariants*. Furthermore for any prime p and k arbitrary, theorem 41 implies

$$f \cong f_p \equiv g_{0p} + pg_{1p} + p^2 g_{2p} + \cdots + p^r g_{rp} \pmod{p^k}$$

where g_{ip} has n_{ip} variables and unit determinant in $R(p)$. We call g_{ip} (except for a square unit factor), n_{ip}, $c_p(g_{ip})$ and, when $p = 2$ the properly primitiveness or improperly primitiveness of g_{ip}, the *ring invariants*. If for a particular prime p the field and ring invariants are equal, theorem 36 shows that the two forms are equivalent in $R(p)$. (For $p = 2$ this condition is not necessary.) We now state and prove

THEOREM 46. *Given a set of field invariants satisfying the conditions of theorem 29 and a set of ring invariants satisfying* *for each g_{ip} the conditions 3 and 4 of theorem 29 and consistent with the field invariants, there is a form with integral coefficients having the given ring and field invariants. By "consistency" of the field and ring invariants is meant the satisfaction of the following conditions for all p:*

1. $n_{0p} + n_{1p} + \cdots + n_{rp} = n$.
2. $\sum_{i=1}^{r} in_{ip} = n_p$, the highest power of p in d.
3. $\Pi_{i=0}^{r} \mid g_{ip} \mid \cong d/p^{n_p}$ in $R(p)$.
4. $c_p(f) = c_p(g_{0p} + pg_{1p} + \cdots + p^r g_{rp})$.
5. *If for any i, g_{i2} is improperly primitive, n_{i2} is even and* $\mid g_{i2} \mid \equiv (-1)^{n_{i2}/2} \pmod 4$.

To prove our theorem we first from theorem 29 construct a form f with integral coefficients and having the given field invariants. Then, for some prime p dividing d, by theorem 29, construct an integral form h_{ip} in n_{ip} variables such that $\mid h_{ip} \mid = \mid g_{ip} \mid$, $c_p(h_{ip}) = c_p(g_{ip})$ and for $p = 2$, h_{ip} properly or improperly primitive as desired. Then, since the ring invariants are consistent

$$f \sim g_{0p} + pg_{1p} + \cdots + p^r g_{rp} \text{ in } F(p).$$

Hence by the lemma there is a transformation T of determinant 1, with rational elements whose denominators are powers of p, taking f into an integral form $f_p \equiv g'_{0p} + pg_{1p} + \cdots + p^r g_{rp} \pmod{p^k}$ for a preassigned arbitrary k and where g'_{0p} has the same field and ring invariants as g_{0p}. This transformation T is in $R(q)$ for all primes q different from p and hence alters no ring or field invariants for any prime q different from p. This process may be carried through for all primes p dividing $2d$ to give the desired result.

Chapter VI

REPRESENTATIONS BY FORMS

35. Introduction. If f and g are forms with integer coefficients and n and m variables respectively ($n > m$) the results of the last chapter give us methods of finding whether or not some form in the genus of f represents g integrally. Corollary 44a or 44b may be used to show that the existence of such a representation depends on the solvability of the congruence $f \equiv g \pmod{8 \mid g \mid P}$ where P is the product of the odd primes in $\mid g \mid \cdot \mid f \mid$, or on the existence of representations in $R(p)$ for all p dividing $2 \mid f \mid \mid g \mid$. When there is only one class in the genus of f, the same criteria serve to determine the existence of representations of g by the form f. However, when there is more than one class in the genus except for certain very special cases and asymptotic results there are no known criteria for existence of representations.

When it comes to determining the number of representations of g by f, known results, except for those in section 38 below, depend on analytic theory which is beyond the scope of this book. However we shall describe such conclusions.

For the case $m = n$ two fundamental problems arise. First, the question of equivalence cannot in general be elegantly resolved in the ring of rational integers. Faced with such a problem one would first test for semi-equivalence by methods of the previous chapter; then employ a reduced form such as is described in theorem 23. Except for binary and ternary forms where unique reduced forms have been found (see the following chapters) luck and perseverance seem to be the best tools beyond those mentioned in the previous sentence. Second, finding the number of transformations taking a form into an equivalent form is equivalent (see section 30) to finding the number of automorphs of a form. Unless the form is definite, the number of automorphs is infinite. For positive forms general results are in terms of the so-called mass function described in the next section which involves all classes in the genus.

36. **Siegel's representation function.** In view of the remarks of the previous section it is natural to suppose that any expression for the number of representations should involve all forms of the genus f and would depend on the number of solutions of certain congruences. This is in accord with Siegel's results (see references 10, 11) which we first describe for f and g positive forms. Let S and T be the matrices of f and g respectively and $E(S)$ the number of automorphs of S, that is, the number of integral matrices taking S into itself. We denote by $A(S, T)$ the number of representations of T by S and write

$$M(S, T) = \Sigma \frac{A(S_k, T)}{E(S_k)}, \qquad M(S) = \Sigma \frac{1}{E(S_k)}$$

where the sums are over all classes in the genus of f. $A_q(S, T)$ stands for the number of distinct solutions (mod q) of $X^T SX \equiv T \pmod{q}$. Then Siegel's result is the formula

$$A_0(S, T) = \frac{M(S, T)}{M(S)} = \tau \lim_{q \to \infty} A_q(S, T) q^v \tag{25}$$

where $v = m(m + 1)/2 - mn$ and τ is a constant depending only on m, n, $|S|$, $|T|$. Siegel, and others, showed that if p^b is the highest power of p in $2|T|$, $A_q(S, T)q^v$ has, for

$q = p^a$ a value independent of a for all $a > 2b$. Hence in the right side of equation (25) the limit may be disregarded if q is taken sufficiently large; in fact, see below, it may be replaced by a product over all primes p. Notice that in cases of genera of one class, the left side of (25) reduces to $A(S, T)$, the number of representations of g by f. The expression $M(S)$ is called the *mass* of the genus. Minkowski found its value except for an error of a power of 2. Siegel's formula is

$$M(S) = \frac{2\Gamma(1/2)\Gamma(2/2)\cdots\Gamma(n/2)\, d^{(n+1)/2}}{\pi^{n(n+1)/4}\Pi\, \alpha_p(S, S)}, \qquad n > 1$$

where Γ denotes the gamma function and $\alpha_p(S, S) = \frac{1}{2}q^w A_q(S, S)$ with $w = -(n - 1)n/2$ and $q = p^a$ for sufficiently large a as above.

If f is a positive form and g is the number N, Siegel's formula reduces to the following

$$(26) \qquad A_0(S,N) = \frac{\tau N^{\frac{1}{2}n-1}\, \pi^{n/2}}{\Gamma(n/2)\cdot\sqrt{d}}\, \Pi\, \frac{A_q(S, N)}{q^{n-1}}$$

where d is the determinant of S, τ is 1 or 1/2 according as $n > 2$ or $n = 2$ and the product ranges over all $q = p^a$, where p is a prime and, as above, $a > 2b$ with b the highest power of p dividing $2d$.

Since an indefinite form has an infinite number of automorphs the number of representations is infinite if not zero. Hence the left side of (25) is replaced by a certain limiting quotient of volumes while the right remains unaltered. Extensions of these results are possible even for forms with algebraic coefficients.[13]

37. **Asymptotic results.** Known asymptotic results have so far been confined to representations of numbers by forms. Tartakowski (reference 15) has shown that if $f \equiv N$

(mod M) is solvable for an arbitrary M, that is, if f represents N in $R(p)$ for all p, there is a constant G depending on n such that $N > G$ implies that f represents N provided f has more than four variables and is a positive form. For quaternary and ternary forms certain other restrictions must be imposed on N (cf. the examples in the next section).

Kloosterman (reference 6) proved a result for forms without cross products which Ross and Pall (references 7, 9) extended to all other forms and which, in Siegel's notation, is the following:

$$/A(S, N) - W < B \cdot N^{n/2-1-1/18+\epsilon}$$

for some number B independent of n and N, where W is the quantity on the right side of equation (26) and ϵ is an arbitrary positive number, while f is a positive n-ary form with $n \geqq 4$ and having integer coefficients while N is a positive integer (S being the matrix of f).

Suppose

$$f = \sum_{i=1}^{n} a_i x_i^2, \qquad h = \sum_{i=1}^{n} |a_i| x_i^2, \; a_i \text{ non-zero integers.}$$

Siegel then dealt with the expression

$$A(\epsilon) = \Sigma \exp(-\pi\epsilon h)$$

the sum being extended over all integral solutions of $f = N$. Using the deep analytical methods of Siegel, Mary Dolciani[2] proved the following results:

Let f be an n-ary form and N an integer such that $f = N$ is solvable in $R(p)$ for all prime divisors of $2d$, with $d = |f|$, and for the infinite prime. Then

1. If f is indefinite and $n > 4$,

$$\lim_{\epsilon \to 0^+} \epsilon^{n/2-1} A(\epsilon) \text{ exists and is positive,}$$

2. If f is indefinite and $n = 4$,

$$\lim_{\epsilon \to 0+} \epsilon A(\epsilon) \text{ exists and is positive,}$$

unless $N = 0$ and d is a square in which case

$$\lim_{\epsilon \to 0+} (\epsilon/\log \epsilon^{-1}) A(\epsilon) \text{ exists and is positive.}$$

3. If f is positive and $n > 4$

$$\lim_{N \to \infty} N^{1-n/2} A(N^{-1}) \text{ is positive.}$$

4. If f is positive and $n = 4$, then

$$\lim_{N \to \infty} \frac{\log N}{N} A(N^{-1}) \text{ is positive}$$

provided, for each prime dividing $2d$, one of the following conditions holds:

a) f represents N primitively in $R(p)$.

b) There is a number B independent of N such that the power of p in every N of the limit is less than B.

c) f represents zero in $R(p)$.

These results carry over to forms with cross products. They give an estimate of the number of representations. Notice that in each case the existence of the limit implies that f represents N since for $N \neq 0$, $A(\epsilon) > 0$ if and only if f represents N, while for $N = 0$, $A(\epsilon) > 1$ if and only if f is a zero form. In other words, if N is a number represented by f in $R(p)$ for $p = \infty$ and all primes dividing $2d$, $f = N$ is solvable if f is indefinite with $n \geqq 4$ or if N and f are positive, $n > 4$ and N "sufficiently large". If f is positive, $n = 4$ and $N > 0$ the same result holds if the power of p dividing N is bounded for every p in $2d$. Kloosterman's results give the same implications for positive forms. All these results are obtained by deep analytical methods.

One of the outstanding problems in the theory of quadratic forms is the extension of such results as the above to ternary forms. For instance, the following example shows that the above result for positive quaternary forms does not hold for positive ternary forms. It can be shown that the forms $f = x_1^2 + x_2^2 + 16x_3^2$ and $g = 2x_1^2 + 2x_2^2 + 5x_3^2 - 2x_2x_3 - 2x_1x_3$ are representatives of the only classes in the genus of f. Pall and the author (see reference 8) have shown that the number of representations by f of any number $N \equiv 1 \pmod 8$ is the same as the number of representations by g except that g represents no odd perfect square. A few other like results hold.

38. **A representation function.** The representation function developed in the remainder of this chapter does not depend on complex analysis. Furthermore it is equally valid in the major part of the theory for definite or indefinite forms. On the other hand its results are not very specific except for the cases $m = n - 1$ and $m = 1$, where f and g have n and m variables, respectively. We say that X is a *primitive representation* of B by A if $X^TAX = B$ with 1 the g.c.d. of the m-rowed minors of X and deal primarily with primitive representations. We assume throughout this section that A and B are non-singular.

We have seen that the number of representations is intimately connected with the automorphs of a form since if T is a solution of $X^TAX = B$, then PT is also a solution for P any automorph of A. In the remainder of this chapter we conform to classical practice and consider a *transformation to be an automorph if it takes the form into itself, has integral elements and determinant* $+1$, though such conformity is not really necessary. Also in this section, we require *a unimodular matrix to have determinant* $+1$.

We then call two solutions T_1 and T_2 of $X^T AX = B$ *essentially distinct* if there is no automorph P (in the above sense) of A such that $T_1 = PT_2$. If such an automorph exists, we call the solutions *essentially equal*. Not only does this seem a natural definition but it has the advantage of allowing us to deal simultaneously with positive and indefinite forms, for though the number of automorphs of the latter is infinite, the number of essentially distinct solutions of $X^T AX = B$ we show to be finite. If then we let $N(A, B)$ denote the number of essentially distinct primitive solutions of $X^T AX = B$, noting that multiplication by an automorph takes a primitive solution into a primitive solution, we define our number of representations functions to be

$$M(d, B) = \Sigma\, N(A_k, B), \qquad G(A, B) = \Sigma\, N(A_k, B)$$

where the first sum is over a set of forms A_k of determinant d, the second is over a set of forms of the genus of A, and in both cases each class of forms has one and only one representative. It is this function which we seek to evaluate.

We shall find useful the following

Lemma 13. *If T is a primitive n by m matrix $(n > m)$ with integral elements and T_0 and T_0' are two n by $n - m$ matrices with integral elements such that the matrices $(T\ T_0)$ and $(T\ T_0')$ are unimodular, then there exists an m by $n - m$ matrix R with integer elements and a unimodular matrix S with $n - m$ rows such that $T_0' = TR + T_0 S$. Conversely, for any integral R and unimodular S, $(T\ T_0)$ unimodular implies that $(T\ TR + T_0 S)$ is unimodular.*

To prove this first write

$$(T\ T_0)^I (T\ T_0') = \begin{bmatrix} R_1 & R \\ S_2 & S \end{bmatrix}$$

which is unimodular if the matrices on the left are unimodular. Then multiplication of the above equation on the left by $(T\ T_0)$ yields $TR + T_0S = T_0'$. Furthermore S is unimodular since if we let

$$\begin{bmatrix} \overline{T} \\ \overline{T}_0 \end{bmatrix} = (T\ T_0)^I$$

it follows that $\overline{T} \cdot T = I$, $\overline{T}_0 \cdot T_0 = I$, $\overline{T} \cdot T_0 = (0)$, $\overline{T}_0 \cdot T = (0)$ and hence

$$(T\ T_0)^I(T\ T_0') = \begin{bmatrix} \overline{T} \\ \\ \overline{T}_0 \end{bmatrix} [T\ T_0'] = \begin{bmatrix} \overline{T}T & \overline{T}T_0' \\ \\ \overline{T}_0T & \overline{T}_0T_0' \end{bmatrix} = \begin{bmatrix} I & \overline{T}T_0' \\ \\ 0 & \overline{T}_0T_0' \end{bmatrix}$$

where I is used to stand for identity matrices of the proper sizes and 0 for zero matrices. Hence $S = \overline{T}_0T_0'$ and has determinant $+1$.

The converse statement is shown by the fact that

$$[T\ TR + T_0S]\begin{bmatrix} I & -RS^I \\ 0 & S^I \end{bmatrix} = [T\ T_0].$$

If now T is a primitive solution of $X^TAX = B$, we can by lemma 6 find an n by $n - m$ matrix T such that $(T\ T_0)$ is unimodular and takes A into $\begin{bmatrix} B & C \\ C^T & D \end{bmatrix}$ where $C = T^TAT_0$ and $D = T_0^TAT_0$.

It is useful also to consider the adjoint matrices. If X is a submatrix of A we denote by $\bar{X}$ that matrix obtained from X by replacing each element by its cofactor in A and by the same token $\bar{A}$ is the *transpose of the adjoint* of A. Hence if A is symmetric, $\bar{A}$ is its adjoint. Then

$$A' = \begin{bmatrix} B & C \\ C^T & D \end{bmatrix} \text{ implies } \bar{A}' = \begin{bmatrix} \bar{B} & \bar{C} \\ \bar{C}^T & \bar{D} \end{bmatrix}$$

and if $|A'| = d$, then the equation $A'\bar{A} = dI$ implies $B\bar{B} + C\bar{C}^T = dI$, $C^T\bar{C} + D\bar{D} = dI$ and $B\bar{C} + C\bar{D} = (0)$.

If we solve the last equation for $\bar{C}$ and substitute in the second, we get

$$(27) \qquad (-C^T B^I C + D)\bar{D} = dI.$$

Hence $|\bar{D}| = d^{n-m}|D - C^T B^I C|^{-1}$. But the transformation

$$\begin{bmatrix} I & -B^I C \\ 0 & I \end{bmatrix}$$

of determinant 1 takes A' into $B \dotplus (D - C^T B^I C)$, where $\dotplus$ denotes direct sum. This shows that $|D - C^T B^I C| q = d$, where $q = |B|$ and

$$|\bar{D}| = qd^{n-m-1}.$$

Furthermore if we define E to be the integral matrix $q(D - C^T B^I C) = qD - C^T(\text{Adj } B)C$, we have $|E| = dq^{n-m-1}$.

Then, continuing the above notation, we have $(T\ T_0)^I = (\bar{T}\ \bar{T}_0)^T$ and $(T\ T_0)^T A (T\ T_0) = A'$ implies $(T\ T_0)^I \bar{A} (T\ T_0)^{IT} = \bar{A}'$, that is, $(\bar{T}\ \bar{T}_0)^T \bar{A} (\bar{T}\ \bar{T}_0) = \bar{A}'$ and hence $\bar{T}_0$ takes $\bar{A}$ into $\bar{D}$.

The lemma above shows that if $(T\ T_0')$ and $(T\ T_0)$ are unimodular then $T_0' = TR + T_0 S$ for properly chosen integral matrices R and S with the latter unimodular. We now show that replacing T_0 by $TR + T_0 S$ replaces C by $BR + CS$, $\bar{D}$ by $S^I \bar{D} S^{IT}$ and hence E by $S^T E S$. The equations $T^T A T_0' = T^T A T R + T^T A T_0 S = BR + CS$ establish the first part of the statement. The second part follows from the fact that

$$[T\ T_0]\begin{bmatrix} I & R \\ 0 & S \end{bmatrix} = [T\ TR + T_0 S] = [T\ T_0']$$

which implies that

$$[T\ T_0']^I = \begin{bmatrix} I & -RS^I \\ 0 & S^I \end{bmatrix}[T\ T_0]^I = \begin{bmatrix} I & -RS^I \\ 0 & S^I \end{bmatrix}\begin{bmatrix} \bar{T}^T \\ \bar{T}_0^T \end{bmatrix} = \begin{bmatrix} \bar{T}'^T \\ \bar{T}_0'^T \end{bmatrix}.$$

Now let $\mathfrak{S}$ be a set of forms in $n - m$ variables of determinant dq^{n-m-1} and having the following properties:

1) No two forms of $\mathfrak{S}$ are equivalent,
2) If E is a form of $\mathfrak{S}$ there are integral matrics D and C such that $E = qD - C^T(\text{Adj } B)C$,
3) There is no larger set having properties 1 and 2.

Then from the above we see that if T is a primitive representation of B by A and T_0 is so chosen that $(T\ T_0)$ is unimodular we can choose a unimodular matrix S such that $E = qD - C^T(\text{Adj } B)C$ is the matrix of a form of set $\mathfrak{S}$ where $C = T^T A T_0 S$ and $D = S^T T_0^T A T_0 S$. Furthermore $(T\ T_0)$ and $(T\ BR + T_0 S)$ yield the same E if and only if S is an automorph of E. Hence we see that if C_1 and C_2 are two matrices $T^T A T_0$ and $T^T A T_0'$ for T_0 and T_0' yielding the same chosen E in $\mathfrak{S}$, then the equation $C_2 = BR + C_1 Q$ holds for some automorph Q of E and matrix R (with integral elements).

Given any E of the set $\mathfrak{S}$ and C_1 and C_2 two solutions of the congruence $E \equiv -C^T(\text{Adj } B)C \pmod{q}$, we call C_1 and C_2 *essentially equal* if there is an integral matrix R and an automorph Q_E of E such that $C_2 = BR + C_1 Q_E$. If C_1 and C_2 are not essentially equal we call them *essentially distinct*. (Notice that if $q = \pm 1$, all C are essentially equal.) Thus we have shown that *to any primitive representation T of B by A corresponds a unique form E of $\mathfrak{S}$ and an essentially unique matrix C such that*

$$\begin{bmatrix} B & C \\ C^T & D \end{bmatrix}$$

has determinant d and $E = qD - C^T(\text{Adj } B)C$. Since, if we replace T by PT for any automorph P of A, we leave B, C, D and E unaltered and the above italicized statement can be made for any two essentially equal T.

On the other hand we can now show that if $(T_i\ T_{i0})$ take A into

$$\begin{bmatrix} B & C_i \\ C_i^T & D_i \end{bmatrix} \tag{28}$$

where $E = qD_i - C_i^T(\text{Adj } B)C_i$ for $i = 1, 2$ and $C_2 = BR + C_1Q_E$ for some integral matrix R and automorph Q_E of E, then there is an automorph P of A such that $T_2 = PT_1$. Note that the relationship between C_1 and C_2 shows that the transformation

$$\begin{bmatrix} I & R \\ 0 & Q_E \end{bmatrix}$$

takes

$$\begin{bmatrix} B & C_1 \\ C_1^T & D_1 \end{bmatrix} \text{ into } \begin{bmatrix} B & C_2 \\ C_2^T & L \end{bmatrix}$$

where $L = R^TBR + Q_E^TC_1R + R^TC_1Q_E + Q_E^TD_1Q_E$. Then, using the facts that $R = B^I(C_2 - C_1Q_E)$ and $Q_E^T(D_1 - C_1^TB^IC_1)Q_E = D_2 - C_2^TB^IC_2$, we find that $L = D_2$. Hence

$$[T_1\ T_{10}] \begin{bmatrix} I & R \\ O & Q_E \end{bmatrix} [T_2\ T_{20}]^I = P^I$$

which is an automorph of A and $T_2 = PT_1$.

Thus we have shown that *two representations of B by A yield the same class E of $\mathfrak{G}$ and two essentially equal matrices C_i if and only if the two representations are themselves essentially equal:* Since, for each E of $\mathfrak{G}$, every solution of

$$E \equiv -X^T(\text{Adj } B)X \pmod{q} \tag{29}$$

yields a matrix

$$\begin{bmatrix} B & C \\ C^T & D \end{bmatrix}$$

with $E = qD - C^T(\text{Adj } B)C$ and determinant d, we have proved

THEOREM 47a. *The function*

$$M(d, B) = \Sigma\, P(d, B, E_i)$$

where the sum is over all forms E_i of $\mathfrak{G}$ and $P(d, B, E)$ denotes the number of essentially distinct solutions C of (29). *If $q = \pm 1$, $P(d, B, E_i) = 1$.*

If q is prime to $2d$ (this includes $q = \pm 1$), if $E_1 \text{ v } E_2$ and C_i are any solutions of $E_i \equiv -X^T(\text{Adj } B)X \pmod{q}$ with $E_i = qD_i - C_i^T(\text{Adj } B)C_i$ and $|E_i| = dq^{n-m-1}$ then we can show that the matrices A_i of form (28) for $i = 1, 2$, are semi-equivalent. For let W be a transformation with rational elements whose denominators are prime to $2qd$ and taking E_1 into E_2. Then the transformation

$$\begin{bmatrix} I & -B^I C_1 \\ 0 & I \end{bmatrix} \begin{bmatrix} I & 0 \\ 0 & W \end{bmatrix} \begin{bmatrix} I & -B^I C_2 \\ 0 & I \end{bmatrix}^I$$

takes A_1 into A_2 and has denominators prime to $2d$ where $d = |A_i|$. Furthermore theorem 40 shows that $E_1 \text{ v } E_2$ implies that for an arbitrary integer ∇, there is a form E_1' equivalent to E_1 such that $E_1' \equiv E_2 \pmod{\nabla}$; hence if (29) is solvable for one form E it is solvable for every form in its genus. On the other hand as we shall see later, $A_1 \text{ v } A_2$ does not imply that $E_1 \text{ v } E_2$.

Thus we have

THEOREM 47b. *If q is prime to $2d$ (including $q = \pm 1$) then*

$$G(A, B) = \Sigma\, P(d, B, E_i)$$

where the sum is over all forms E_i in $\mathfrak{G}$ such that for some solution C_i of (29) *with $E = E_i$, the matrix* (28) *is in the genus of A and D is defined by $E_i = qD_i - C_i^T(\text{Adj } B)C_i$.*

Furthermore, if $P(d, B, E_i)$ does not vanish for one value of i, it is different from zero for all forms E_i of the same genus. Also, in virtue of the result cited above, the number of solutions (mod q) *of* (29) *is the same for all forms in the genus of E.*

However, $P(d, B, E_1)$ may be different from $P(d, B, E_2)$ since the automorphs of E_1 and E_2 are also involved and may be different (mod q). For $q = \pm 1$ *the argument used for $m = 1$ below shows that $A_1 \text{ v } A_2$ implies $E_1 \text{ v } E_2$ unless one of the E_i represents a unit and the other does not; furthermore at most two genera of E_i can occur.*

We now proceed to make our results more specific for the cases $m = n - 1$ and $m = 1$.

The case $m = n - 1$.

Here there is just one form E, namely $E = d$. It has one automorph, the number 1. Hence $M(d, B) = P(d, B, d)$ which is the number of matrices C distinct (mod B) such that

$$(30) \qquad d + C^T(\text{Adj } B)C \equiv 0 \pmod{q}.$$

We may find as follows the number of incongruent (mod q) matrices BR as R ranges over all matrices $(r_1, r_2, \cdots, r_{n-1})^T$ (mod q). There exist unimodular matrices U and V such that $UBV = B' \equiv b_1 \dotplus b_2 \dotplus \cdots \dotplus b_{n-1} \pmod{q^2}$. Then, letting $R' = V^I R$ we seek the number of incongruent matrices $B'R' = (b_1' u_1 r_1', b_2' u_2 r_2', \cdots, b_{n-1}' u_{n-1} r_{n-1}')$ where $R' = (r_1', r_2', \cdots, r_{n-1}')$ and $b_i = b_i' u_i$ with $\Pi b_i' = q$ and $\Pi u_i \equiv 1 \pmod{q}$. Then $b_i' u_i r_i'$ takes on q/b_i' distinct values (mod q) as r_i' ranges over a complete set of residues (mod q). Hence $B'R'$ takes on q^{n-2} distinct values. Thus we have

THEOREM 48. *The function $M(d, B)$ is equal to the product of q^{2-n} and the number of matrices C distinct* (mod q) *satisfying* (30).

If, in particular, q is prime to $2d$, $M(d, B) = G(A, B)$ for all matrices A representing B are of the same genus since

$$V = \begin{bmatrix} I & B^I(C_2 - C_1) \\ 0 & I \end{bmatrix}$$

is a matrix taking A_1 into A_2 where A_i have the form (28) and V has rational elements with denominators prime to $2d$. Furthermore if $q = \pm 1$, $G(A, B) = 1$.

In the next chapter we shall consider in more detail the case $n = 2$, $m = 1$.

The case $m = 1$.

Here $B = q$, $E = qD - C^T C$ and, from theorem 47a,

$$M(d, q) = \Sigma\, P(d, q, E_i)$$

where the sum is over all forms E_i of $\mathfrak{G}$ and $P(d, q, E_i)$ denotes the number of essentially distinct solutions C of

$$(31) \qquad E_i \equiv -X^T X \pmod{q}.$$

Considerable simplification occurs when we impose the condition that q shall be prime to d. The case of $q = \pm 1$ is included almost trivially in what follows. Then if C is any matrix C_i in (28) we see that the g.c.d. of its elements is prime to q since otherwise the determinant of A would be divisible by a factor of q. Hence, since C may be replaced by any matrix congruent to it (mod q), we may assume 1 to be the g.c.d. of its elements and hence the existence of a unimodular matrix U such that $CU = C_0 = (1, 0, \cdots, 0)$. Thus if C is a solution of (31), $U^T E_i U \equiv -C_0^T C_0 \pmod{q}$ for all E_i in $\mathfrak{G}$ and we may consider the E_i replaced by $U^T E_i U$. Hence $P(d, q, E_i)$ is the number of essentially distinct matrices C such that $C^T C \equiv C_0^T C_0 \pmod{q}$.

We recall that C_1 and C_2 are essentially distinct if

there is no automorph Q_E of E and integral matrix R such that $C_2 = qR + C_1Q_E$, that is, if there is no automorph Q_E of E such that $C_2 \equiv C_1Q_E \pmod q$. Hence to simplify our expression for $M(d, q)$ we need more information about the automorphs (mod q) of E.

Notice that $E \equiv -C_0^TC_0 \pmod q$ which implies that $Q^TC_0^TC_0Q \equiv C_0^TC_0 \pmod q$ for any automorph Q of E. Hence the first row of Q is congruent to $(w, 0, \cdots, 0)$ (mod q) with $w^2 \equiv 1 \pmod q$. On the other hand, any solution $C = (c_1, c_2, \cdots, c_{n-1})$ of $X^TX \equiv C_0^TC_0 \pmod q$ must have elements satisfying the conditions $c_1^2 \equiv 1$, $c_i \equiv 0 \pmod q$ for $i = 2, 3, \cdots, n-1$ and the number of such distinct C (mod q) is $2^{\mu(q)}$ where $\mu(q)$ is the number of distinct prime factors of q increased by 1 if $q \equiv 0 \pmod 8$ or decreased by 1 if $q \equiv 2 \pmod 4$. Thus, if λ_E is the number of distinct values (mod q) of the leading elements (that is, elements in the upper left corners) of the automorphs of $E \equiv -C_0^TC_0 \pmod q$ we see that the number of essentially distinct C satisfying (31) is $2^{\mu(q)}/\lambda_E$ and hence we have

THEOREM 49a. *If q is prime to d*

$$M(d, q) = 2^{\mu(q)}\Sigma\, \lambda_E^{-1}$$

where the sum ranges over all classes E of determinant dq^{n-2} such that $E \equiv -C_0^TC_0 \pmod q$ and $\mu(q)$ is the number of distinct prime factors of q increased by 1 if $q \equiv 0 \pmod 8$ or decreased by 1 if $q \equiv 2 \pmod 4$.

The results of the first part of this section show the truth of the first sentence of the theorem below and the remaining discussion of this section suffices to prove the rest of

THEOREM 49b. *If q is prime to $2d$*

$$G(A, q) = 2^{\mu(q)}\Sigma\, \lambda_E^{-1} \tag{32}$$

where the sum is over those classes of matrices E in genera corresponding to the genus of A. At most two genera of matrices E occur (for more details see below) and, if a class E occurs, all classes in the genus occur.

In both theorems $\lambda_E = 2$ if q is a prime and $\lambda_E = 1$ if $q = \pm 1$.

We can show that if q is prime to $2d$ and A_i, written in form (28), are in the same genus, then the corresponding E_i are equivalent in the ring $R(p)$ of p-adic integers for all odd primes p and $p = \infty$; furthermore $E_1 \cong E_2$ in $R(2)$ if and only if the forms having matrices E_1 and E_2 both represent odd numbers or neither represent odd numbers. In other words, from theorem 40 we must show that A_1 and A_2 equivalent in $R(p)$ implies E_1 and E_2 are equivalent in $R(p)$ for all $R(p)$ except $R(2)$ and that A_1 and A_2 equivalent in $R(2)$ implies E_1 and E_2 are equivalent in $R(2)$ if their forms both represent odd numbers or neither represent odd numbers. The chapter on ternary forms shows that there are cases in which for a given genus of A, there is an E_1 representing a unit and an E_2 which does not; hence this exception cannot be avoided.

First suppose p is not a divisor of $2q$. Then the transformations

$$\begin{bmatrix} I & -C_i q^{-1} \\ 0 & I \end{bmatrix}$$

have elements in $R(p)$ and take A_i into $q \dotplus (D_i - C_i^T C_i q^{-1}) = q \dotplus E_i q^{-1}$. Hence theorem 37 shows that $E_1 \cong E_2$ in $R(p)$. The same argument can be used for $p = \infty$.

Second if $p = 2$, theorem 38 shows that q odd and $q \dotplus E_1 q^{-1} = q \dotplus E_2 q^{-1}$ in $R(2)$ implies $E_1 \cong E_2$ in $R(2)$ if and only if both forms represent odd numbers or both represent only even numbers.

Finally if p divides q we have seen above that there is a unimodular matrix U_i such that $C_i U_i = C_0$ where $C_0 = (1, 0, \cdots, 0)$ and $i = 1, 2$. Then the transformation $1 \dotplus U_i$ takes

$$\begin{bmatrix} q & C_i \\ C_i^T & D_i \end{bmatrix} \text{ into } \begin{bmatrix} q & C_0 \\ C_0^T & D_i' \end{bmatrix}$$

where $D_i' = U_i^T D_i U_i$. We need to show that $D_1' - C_0^T C_0 q^{-1} \cong D_2' - C_0^T C_0 q^{-1}$ in $R(p)$. Let $D_1' = (d_{ij})$ and $D_2' = (e_{ij})$. If we multiply the first column of the matrix $qD_1' - C_0^T C_0$ by $-qd_{1i}/(qd_{11} - 1)$, add it to the i-th for $i = 2, 3, \cdots, n - 1$ and do the same for the rows we get $qD_1' - C_0^T C_0 \cong (qd_{11} - 1) \dotplus qD_0$ in $R(p)$ where D_0 is a matrix in $R(p)$ with determinant $d(qd_{11} - 1)^{-1}$. Similarly $qD_2' - C_0^T C_0$ is equivalent in $R(p)$ to $(qe_{11} - 1) \dotplus qE_0$. But the congruence $qd_{11} - 1 \equiv qe_{11} - 1 \pmod{p}$ implies that there is an element u of $R(p)$ such that $qd_{11} - 1 = u^2(qe_{11} - 1)$. Also $| D_0 | = d(qd_{11} - 1)^{-1} = u^{-2} | E_0 |$ with u prime to p implies from corollary 36b that $D_0 \cong E_0$ in $R(p)$. Hence $(qe_{11} - 1) \dotplus qE_0 \cong (qd_{11} - 1) \dotplus qD_0$ in $R(p)$ and we have proved our result.

It perhaps should be remarked that the restriction that representations be primitive may be removed if we replace $M(d, q)$ by $\Sigma\, M(d, q/r^2)$ the sum being over all integers r whose squares divide q. A similar remark may be made about $G(A, q)$.

In the chapter on ternary forms we shall make special applications of these results.

Chapter VII

BINARY FORMS

39. **Introduction.** Most of the results in this chapter are classical—some dating back to the time of Gauss and earlier—and can be derived independently of the previous general theory. But viewing the binary forms as special cases of our previous results illuminates the general theory on the one hand and economizes labor on the other. Furthermore certain problems, such as the determination of all automorphs, inaccessible in the general case, can be completely solved for binary forms.

Since much of the theory of binary forms was developed in advance of the general theory there is a wide divergence in the use of the term "determinant" as applied to a form. Gauss wrote the binary form as $f = ax^2 + 2bxy + cy^2$ and defined the determinant of f to be $b^2 - ac$. Kronecker preferred $f = ax^2 + bxy + cy^2$ and called $b^2 - 4ac$ its determinant. These expressions or their negatives have been variously referred to as the "discriminant" of the form. The confusion of terminology is so great that, in reading the literature, one must take great care to inform himself of the meaning of the author. We shall in this book make a clean break with tradition and define the determinant of a binary form just as it was defined for forms in more variables. That is, the determinant of $ax^2 + 2b_0xy + cy^2$ shall be $ac - b_0^2$ and that of $ax^2 + bxy + cy^2$ shall be $ac - b^2/4$.

In this chapter we shall confine attention to forms $f = ax^2 + bxy + cy^2$ where a, b and c are integers and shall reserve the letters a, b, c for the coefficients, denoting on occasion half the middle coefficient by b_0. It will be taken for granted that the determinant of the form is not zero. Notice that it may be one-fourth of an odd integer if the middle coefficient is odd.

40. Automorphs. We have seen that the automorphs of a form provide the key to many problems. Our first concern then is to find the automorphs of binary forms, that is, the transformations with integral elements taking forms into themselves. The determinant of such a transformation must be ± 1. In classical literature automorphs are usually required to have determinant $+1$. When we wish to make such a restriction we shall call it a *proper automorph*. Since $-I$ is an automorph of any form, the distinction between proper automorphs and unrestricted automorphs is important only for forms with an even number of variables.

If $f = ax^2 + bxy + cy^2$ there is no loss of generality in considering f to be *primitive*, that is, 1 to be the g.c.d. of its coefficients since for any integer r the automorphs of f and rf are identical. Furthermore we may say $a \neq 0$ since f represents some non-zero number. Make these restrictions and see that $af = X^2 + dy^2$ where $d = ac - b^2/4$, being the determinant of f, and $X = ax + \frac{1}{2}by$. Hence the transformation

$$T = \begin{bmatrix} a & \frac{1}{2}b \\ 0 & 1 \end{bmatrix}$$

takes F into af where $F = X^2 + dy^2$. Thus if P is an automorph of f, the transformation TPT^I takes F into itself and we shall find the automorphs of f by means of this relationship. Letting $P = (p_{ij})$ and writing $Q = TPT^I$, multiplying matrices shows that

(33)

$$Q = \begin{bmatrix} p_{11} + \frac{1}{2}bp_{21}\,a & -p_{11}b/2 + p_{12}a - b^2p_{21}/4a + bp_{22}/2 \\ p_{21}/a & -bp_{21}/2a + p_{22} \end{bmatrix}$$

If P is a proper automorph of f, we have the three equations

$$(34) \quad ap_{11}^2 + bp_{11}p_{21} + cp_{21}^2 = a$$

$$(35) \quad 2ap_{11}p_{12} + b(p_{11}p_{22} + p_{12}p_{21}) + 2cp_{21}p_{22} = b$$

$$(36) \quad p_{11}p_{22} - p_{12}p_{21} = 1.$$

Substituting (36) into (35) we get

$$(37) \qquad ap_{11}p_{12} + bp_{12}p_{21} + cp_{21}p_{22} = 0.$$

But (34) shows that a divides $p_{21}(bp_{11} + cp_{21})$ and (37) that a divides $p_{21}(bp_{12} + cp_{22})$. Then (36) and the fact that no factor of a divides both b and c shows that a divides p_{21} and enables us to write $p_{21} = ap'_{21}$ where p'_{21} is an integer. This shows that in Q, above, all the elements are integers except that when b is odd some denominators may contain the number 2 or 4.

If we find all the automorphs of F where now we allow some of the elements to be halves and quarters of integers we can obtain by means of (33) the automorphs of f. To this end let

$$Q = \begin{bmatrix} t & v \\ u & w \end{bmatrix}$$

be a transformation with rational elements and determinant ± 1 taking F into itself. Then

$$t^2 + du^2 = 1$$

$$v^2 + dw^2 = d$$

$$tw - uv = \pm 1.$$

If $u \neq 0$ solve the third equation for v and substitute in the second to get $w^2(t^2 + du^2) \mp 2tw + 1 = du^2$. Use the first equation to get $(w \mp t)^2 = 0$ which implies $w = \pm t$ and $v = \mp du$. If $u = 0$ the above equations imply $t = 1$ or

-1, $tw = \pm 1$ and hence $w^2 = 1$ and $v = 0$. Thus in all cases we see that if Q is a transformation with rational elements and determinant ± 1 taking F into itself

$$w = \pm t, \qquad v = \mp du, \tag{38}$$

where the sign is the sign of the determinant of Q.

Then if $P = (p_{ij})$ is an automorph of f, equations (33) and (38) imply

$$p_{11} + bp'_{21}/2 = t, \qquad p'_{21} = u,$$

where t, u is a solution of $x^2 + dy^2 = 1$. This shows that t and u are integers unless b is odd (that is when $4d$ is an odd integer) in which case $2t$ and u are integers of the same parity. Such a solution we call a *semi-integral solution.* Furthermore (38) imply $p_{11} + bp'_{21}/2 = -bp'_{21}/2 + p_{22}$. Hence $p_{11} = t - \frac{1}{2}bu$, $p_{21} = au$, $p_{22} = t + \frac{1}{2}bu$ and (33) with (38) imply $p_{12} = -cu$. Hence we have almost proved

THEOREM 50. *If $f = ax^2 + bxy + cy^2$ is a primitive form with $a \neq 0$ all proper automorphs of f are expressible in the form*

$$P_0 = \begin{bmatrix} t - \frac{1}{2}bu & -cu \\ au & t + \frac{1}{2}bu \end{bmatrix}$$

where t, u is a semi-integral solution of $x^2 + dy^2 = 1$ and $d = ac - b^2/4$.

Now

$$TP_0T^{\text{I}} = \begin{bmatrix} t & -du \\ u & t \end{bmatrix}$$

which takes F into itself. Hence P_0 takes f into itself. Its determinant is $+1$. If b is even the elements of P_0 must be integers. If b is odd $(2t)^2 + (4d)u^2 = 4$ and the fact that $2t$ and u are of the same parity implies that the elements of P_0 are integers. This completes the proof.

If f is a positive form, d > 0 and the only semi-integral solutions of $x^2 + dy^2 = 1$ (with $4d$ an integer congruent to -1 or 0 (mod 4)) are $t = \pm 1$, $u = 0$ except that when $d = 1$ we have also $t = 0$, $u = \pm 1$ and when $d = 3/4$ we have $t = \pm\frac{1}{2}$, $u = 1$ or -1. Hence we have

THEOREM 51a. *If $f = ax^2 + bxy + cy^2$ is a primitive positive form the only proper automorphs of f are $\pm I$ and*

1. *If $d = 1$ the transformations*

$$\begin{bmatrix} \mp\frac{1}{2}b & \mp c \\ \pm a & \pm\frac{1}{2}b \end{bmatrix}.$$

2. *If $d = 3/4$, I and $-I$ multiplied by the following*

$$\begin{bmatrix} (\pm 1 - b)/2 & -c \\ a & (\pm 1 + b)/2 \end{bmatrix}.$$

Notice that if $d = 1$, theorem 23 shows that $f \cong x^2 + y^2$ and if $d = 3/4$, $f \cong x^2 + xy + y^2$.

If $-d$ is a square f is a zero form and $t^2 + du^2 = 1$ with t, u semi-integral implies that $t + \sqrt{-d}\cdot u$ is integral hence $t - \sqrt{-d}\cdot u = t + \sqrt{-d}\cdot u = \pm 1$, that is $t = \pm 1$, $u = 0$ and we have

THEOREM 51b. *If $f = ax^2 + bxy + cy^2$ is a zero form the only proper automorphs of f are $\pm I$. (The restrictions $a \neq 0$ and f primitive are here not necessary.)*

Notice that in theorems 51a and 51b the number of proper automorphs depends only on the determinant of the form. In fact, the number of proper automorphs is 2 except for $d = 1$ and $d = 3/4$ in which cases it is 4 and 6 respectively.

In order to deal with indefinite non-zero forms, which have infinitely many automorphs, we must consider the semi-integral solutions of the equation

$$(39) \quad x^2 + dy^2 = 1, \ -4d \text{ positive and non-square.}$$

This we call the Pell equation even though in classical literature the name is usually applied only if $-d$ is a positive non-square integer. We have found that there is a $1 - 1$ correspondence between the semi-integral solutions of the Pell equation and the proper automorphs of forms of determinant d.

Call t, u a positive solution of (39) if t and u are positive. The positive semi-integral solution T, U is called the *fundamental solution* if $T + U\sqrt{-d} \leqq t + u\sqrt{-d}$ for every positive semi-integral solution t, u. It is interesting to notice that finding the solutions of the Pell equation above is equivalent to finding the units $t + u\sqrt{-d}$ in the quadratic field (see section 42) obtained by adjoining $\sqrt{-4d}$ to the field of rational numbers, a unit being defined as a number $t + u\sqrt{-d}$ such that $(t + u\sqrt{-d})(t - u\sqrt{-d}) = 1$; t, u being semi-integral. The numerical computation of the fundamental solution is best accomplished by means of the continued fraction expansion of $\sqrt{-d}$. We next prove

LEMMA 14. *If $-4d$ is a positive non-square integer congruent to 1 or 0 (mod 4) then every semi-integral positive solution t, u of the Pell equation (39) satisfies the equation*

$$t + u\sqrt{-d} = (T + U\sqrt{-d})^n$$

for a properly chosen positive integer n, where T, U is the fundamental solution.

First we show that if t_i, u_i are semi-integral solutions of the Pell equation for $i = 1, 2$, then t_3, u_3 is a semi-integral solution when t_3 and u_3 are defined by the equation

$$(40) \quad t_3 + u_3\sqrt{-d} = (t_1 + u_1\sqrt{-d})(t_2 + u_2\sqrt{-d}).$$

It is obvious that t_3 and u_3 are integers if d is an integer. If $4d \equiv -1 \pmod 4$, equation (40) implies $t_3 = t_1t_2 - u_1u_2d$, $u_3 = u_1t_2 + t_1u_2$. If t_2 is an integer, u_2 is even and

$2t_3$ and u_3 are integers. If t_1 and t_2 are halves of odd integers, u_1 and u_2 must be odd and again $2t_3$ and u_3 are integers. Furthermore (40) implies the same equality with u_1, u_2, u_3 replaced by their negatives; multiplying the two equations shows that $t_3^2 + du_3^2 = 1$ and $2t_3$ and u_3 are of the same parity, if $4d \equiv -1 \pmod 4$.

Next let T, U be the fundamental solution of (39). Then, from the above,

$$t_m + u_m\sqrt{-d} = (T + U\sqrt{-d})^m \tag{41}$$

defines a semi-integral solution t_m, u_m for every positive integer m. Now let r, s be any positive semi-integral solution. Since $T + U\sqrt{-d} > 1$ there is a non-negative integer m such that

(42)
$$(T + U\sqrt{-d})^m \leqq r + s\sqrt{-d} \leqq (T + U\sqrt{-d})^{m+1}$$

If one of the equalities holds our theorem follows. In the contrary case

$$1 < \frac{r + s\sqrt{-d}}{t_m + u_m\sqrt{-d}} < T + U\sqrt{-d}$$

where t_m, u_m are defined by (41). The fraction is equal to $r_1 + s_1\sqrt{-d}$ where $r_1 = rt_m + su_md$, $s_1 = t_ms - u_mr$ and from the first part of this proof r_1, s_1 is a semi-integral solution of the Pell equation. If we can show it is a positive solution our theorem will be established by the denial of our assertion that T, U is the fundamental solution. Now

$$r^2 + ds^2 = 1 \quad \text{and} \quad t_m^2 + du_m^2 = 1 \tag{43}$$

imply $r > s\sqrt{-d}$, $t_m > u_m\sqrt{-d}$ and hence $r_1 > 0$. Also s_1 has the same sign as $t_m^2s^2 - u_m^2r^2$ which, using (43), on the one hand is equal to $s^2 - u_m^2$ and on the other hand is equal to $(-t_m^2 + r^2)/(-d)$. But (42) implies $t_m + u_m\sqrt{-d}$

$< r + s\sqrt{-d}$ and hence $s^2 > u_m^2$ or $r^2 > t_m^2$ which shows that $s_1 > 0$ and completes our proof.

We can extend this result of the lemma by noting that any semi-integral non-trivial solution of the Pell equation may be made a positive solution by changing the sign of one or more of t and u. Hence for every semi-integral solution we can choose a positive integer n, $v = 1$ or -1, $w = 1$ or -1 so that $vt + wu\sqrt{-d} = (T + U\sqrt{-d})^n$, the trivial solution being given by $n = 0$. Hence we have the corollary below.

COROLLARY. *Every semi-integral solution of the Pell equation is expressible in the form*

$$\pm(T + U\sqrt{-d})^n$$

where n is an integer (positive, negative or zero) and T, U is the fundamental solution.

Now turn to the discussion of proper automorphs of non-zero indefinite primitive forms. Let P_1, P_2 and P_3 be any proper automorphs of such a form with $P_1P_2 = P_3$. There is by theorem 50 a $1 - 1$ correspondence between the automorphs of f and the semi-integral solutions of the Pell equation. If t_i, u_i are the semi-integral solutions of the Pell equation corresponding to the proper automorphs P_i $(i = 1, 2, 3)$ we have, multiplying the matrices P_1 and P_2 and equating the diagonal elements of the product to the corresponding elements of P_3, the equations

$$t_1t_2 - \tfrac{1}{2}b(u_1t_2 + u_2t_1) - du_1u_2 = t_3 - \tfrac{1}{2}bu_3$$

$$t_1t_2 + \tfrac{1}{2}b(u_1t_2 + u_2t_1) - du_1u_2 = t_3 + \tfrac{1}{2}bu_3 .$$

Hence $t_3 = t_1t_2 - du_1u_2$, $u_3 = u_1t_2 + u_2t_1$. In other words, t_3 and u_3 are defined by the equation

$$(44)\quad t_3 + u_3\sqrt{-d} = (t_1 + u_1\sqrt{-d})(t_2 + u_2\sqrt{-d}).$$

Thus the correspondence between proper automorphs and semi-integral solutions of the Pell equation is preserved under multiplication. Then the corollary of the lemma gives us

THEOREM 51c. *If* $f = ax^2 + bxy + cy^2$ *is an indefinite non-zero primitive form all its proper automorphs are included in the formula*

$$\pm \begin{bmatrix} T - \frac{1}{2}bU & -cU \\ aU & T + \frac{1}{2}bU \end{bmatrix}^n$$

where T, U *is the fundamental solution of the Pell equation* $x^2 + dy^2 = 1$ *and* n *is an integer.*

In this connection attention should be called to a result of Siegel (see reference 13) that for any indefinite form f in n variables there is a finite number of automorphs P_i, $i = 1, \cdots, k$ such that every automorph of f is expressible in the form

$$P_1^{a_1} P_2^{a_2} \cdots P_k^{a_k}$$

for integer values of a_i. We have just shown that for $n = 2$, k has the value 2.

So far we have considered only proper automorphs. A form which may be taken into itself by an integral transformation of determinant -1 is called *ambiguous*. If P is such a transformation equations (33) and (38) with the ambiguous sign negative show that $p_{11} = -p_{22}$. Now we prove

THEOREM 52. *If* $f = ax^2 + bxy + cy^2$ *is an ambiguous form there is in the same proper class as* f *a form* $f' = a'x^2 + a'wxy + c'y^2$ *with* $w = 0$ *or* 1 *and having the automorph*

$$\begin{bmatrix} 1 & w \\ 0 & -1 \end{bmatrix}$$

Now $p_{11} = -p_{22}$ and $p_{11}p_{22} - p_{21}p_{12} = -1$ implies

$$p_{12}p_{21} = 1 - p_{11}^2. \tag{45}$$

Postpone the case when $p_{12} = p_{11} - 1 = 0$. Let v be the g.c.d. of p_{12} and $p_{11} - 1$ and let $-p_{12} = vs_{11}$, $p_{11} - 1 = vs_{21}$ define integers s_{11} and s_{21}. Since the latter are relatively prime we can determine integers s_{12} and s_{22} so that

$$S = \begin{bmatrix} s_{11} & s_{12} \\ s_{21} & s_{22} \end{bmatrix}$$

has determinant $+1$. Then (45) implies $p_{21} = v's_{21}$, $p_{11} + 1 = v's_{11}$ for a properly chosen integer v' and we have

$$PS = S\begin{bmatrix} 1 & r \\ 0 & s \end{bmatrix}$$

for integers r and s. If $p_{12} = p_{11} - 1 = 0$, $p_{11} + 1 \neq 0$, choose v' to be the g.c.d. of p_{12} and $p_{11} + 1$ and similarly determine S. Furthermore $|P| = -1$ and $|S| = 1$ implies $s = -1$. If then we take $w = 1$ or 0 according as r is odd or even and

$$R = \begin{bmatrix} 1 & (w - r)/2 \\ 0 & 1 \end{bmatrix}, \qquad P' = \begin{bmatrix} 1 & w \\ 0 & -1 \end{bmatrix}$$

we see that

$$\begin{bmatrix} 1 & r \\ 0 & -1 \end{bmatrix} R = RP'.$$

Thus $P' = (SR)^I P(SR)$ and if A is the matrix of f, $P^T AP = A$ becomes $P'^T A'P' = A'$ where $A' = (SR)^T A(SR)$. Thus A' is the matrix of a form $f' \cong f$ properly which has P' as an automorph. If $f' = a'x^2 + b'xy + c'y^2$ is the form whose matrix is A', the fact that P' is an automorph shows us that $f' = a'(x + wy)^2 + b'(x + wy)(-y) + c'y^2$ and thus $b' = 2a'w - b'$ which shows that $b' = a'$ or 0 according as $w = 1$ or 0. This proves our theorem.

Then all the automorphs of an ambiguous form may be found by putting it into the form f' and multiplying the proper automorphs by I and the automorph of theorem 52. Notice that the forms with $d = 1$ and $d = 3/4$ in theorem 51a are both ambiguous.

41. Representations by binary forms. In this section we consider forms $f = ax^2 + bxy + cy^2$ where b is an even integer and a and c are integers and the solutions of $f = q$, where q is a non-zero integer. Such forms have integral matrices. From theorem 48 immediately follows

THEOREM 53. *The value of $M(d, q)$ is the number of distinct solutions* (mod q) *of the congruence*

$$(46) \qquad x^2 \equiv -d \pmod{q}$$

where, $M(d, q) = \Sigma N(A, q)$ with $N(A, q)$ the number of essentially distinct primitive representations of q by A and the sum is over all classes A of determinant d, the classes not being restricted to be primitive.

Suppose the square of a prime p divides d and q. Then from theorem 41 we may write f in the form $f_0 \equiv a_1x_1^2 + a_2x_2^2 \pmod{p^2}$. If $a_1 \equiv a_2 \equiv 0 \pmod{p}$ there is a 1–1 correspondence between the solutions of $f = q$ and $f' = f/p = q/p$; if $a_1 \not\equiv 0 \pmod{p}$, $f = q$ implies $x_1 = px_1'$ where x_1' is integral and, substituting this value for x_1 we have a 1–1 correspondence between the solutions of $f = q$ and $f' = q/p$. In both cases $|f'| = d/p^2$. By this process we may reduce consideration to the case when q and d have no square factor in common except 1. Now, though theorem 53 seems to give the simplest general result, we can, assuming this reduction to be made, express the results of the above theorem in somewhat more definite form by investigating the number of solutions of (46).

If p^2 divides q but not d and d is divisible by p, there is no solution of $x^2 \equiv -d \pmod{p^2}$ and hence of (46). If $d \equiv q \equiv 0 \pmod{p}$ and p^2 does not divide q, there is one solution of $x^2 \equiv -d \pmod{p}$. If p is an odd prime not dividing d, the number of solutions of (46) for $q = p^a$, $a \geqq 1$ is $1 + (-d \mid p)$ as may be seen from the theory of numbers; if $q = 2^a$ and d odd the number of solutions is c_0 where c_0 is 4 if $a \geqq 3$ with $-d \equiv 1 \pmod{8}$, 2 if $a = 2$ with $-d \equiv 1 \pmod{4}$, 1 if $a = 1$ or 0, 0 otherwise. Thus if q_1, the g.c.d. of q and d, is square-free (46) has no solutions when q/q_1 has a factor in common with d while if q/q_1 is prime to d, (46) has

$$c_0\Pi\{1 + (-d \mid p)\}$$

solutions where the product is over all odd prime factors of q/q_1. Thus we have

THEOREM 54. *If $q = q_1q_2$ where q_1 is the* g.c.d. *of q and d and q_1 is square-free, then*

$$M(d, q) = Lc_0\Pi\{1 + (-d \mid p)\}$$

where the product is over all odd prime factors of q_2, $L = 0$ or 1 according as q_2 has or has not a prime factor in common with d and c_0 has the value 0 except for the following circumstances:

i) $c_0 = 1$ *if* $q \not\equiv 0 \pmod{4}$,
ii) $c_0 = 2$ *if* $q \equiv 4 \pmod{8}$ *and* $-d \equiv 1 \pmod{4}$,
iii) $c_0 = 4$ *if* $q \equiv 0 \pmod{8}$ *and* $-d \equiv 1 \pmod{8}$.

The above product can be written as a sum and incorporated into the following

COROLLARY 54. $M(d, q) = Lc_0\Sigma(-d \mid \mu)$, *the sum being over all square-free odd divisors, μ, of q_2, and L and c_0 having the values above.*

When the g.c.d. of q and $4d$ is square-free we can find expeditiously the value of the representations expression corresponding to $M(d, q)$ with the representations not restricted to be primitive. To this end let $N_0(A, q)$ stand for the number of essentially distinct representations of q by A and $M_0(d, q)$ the sum of $N_0(A, q)$ over all classes of determinant d. Then

$$M_0(d, q) = \sum_{r^2 | q} M(d, q/r^2) = L \sum_{r^2 | q} \sum_{\mu} (-d \mid \mu)$$

the second sum being over all odd divisors μ of q_2/r^2. In the double sum any $(-d \mid \mu)$ occurs to the number of times equal to the number of different values of r such that μr^2 divides q_2. Then, since $(-d \mid \mu) = (-d \mid \mu r^2)$ we have

THEOREM 55. *If q has no square factor greater than* 1 *in common with* $4d$ *and* $q = q_1 q_2$ *where* q_1 *is the* g.c.d. *of* q *and* d,

$$M_0(d, q) = L\Sigma(-d \mid \mu)$$

where L is defined in theorem 54, $M_0(d, q)$ *is the sum of the essentially distinct representations of q by the classes of forms of determinant d and the sum is over all odd divisors of* q_2. *Not all the classes need be primitive.* (*See the note below.*)

NOTE. If q is prime to d the only forms representing q are properly primitive if d is even, $d \equiv 1 \pmod 4$ or $d \equiv 3 \pmod 4$ with q odd; improperly primitive if $d \equiv 3 \pmod 4$ and $q \equiv 2 \pmod 4$. If q is prime to $2d$ the remarks after theorem 48 show that all the classes in $M_0(d, q)$ are of the same genus.

Notice that in the last theorem we required that not only should q have no square factor greater then 1 in common with d but that q should not be divisible by 4. It is possible to reduce the general case to this case by somewhat similar methods to those used to eliminate square factors common to q and d but the details are not especially inspiring. The following two examples will suffice.

EXAMPLE 1. Suppose $d = 28$, $q = 22$. The form f may be considered to be $\equiv a_1x_1^2 + a_2x_2^2 \pmod 4$ and either a_1 and a_2 are both even in which case consideration of $f = 22$ reduces to that of $f/2 = 11$, or a_1 is odd and $f = 22$ implies $x_1 = 2x_1'$ and $f = 22$ is not solvable.

EXAMPLE 2. Suppose $d = 7$, $q = 44$. Then either f is properly primitive and may be taken $\equiv a_1x_1^2 + a_2x_2^2 \pmod 8$ in which case $f = 44$ implies x_1 and x_2 are even and consideration is reduced to $f = 11$, or f is improperly primitive and may be taken to be $\equiv 2x_1x_2 \pmod 8$ when $f = 44$ implies that x_1 or x_2 is even and consideration is reduced to $f = 22$.

Suppose for some integer q, x_0, y_0 is a solution of $f = q$. From the definition of essentially equal solutions and theorem 50 all solutions essentially equal to x_0, y_0 are obtained from this by multiplying the matrix $(x_0\ y_0)^T$ on the left by the matrix

$$Q = \begin{bmatrix} t - \frac{1}{2}bu & -cu \\ au & t + \frac{1}{2}bu \end{bmatrix}.$$

If Q took x_0, y_0 into itself we would have $Q(-I)(x_0\,y_0)^T = 0$ and hence $|Q - I| = 0$, that is $(t - \frac{1}{2}bu - 1)(t + \frac{1}{2}bu - 1) + acu^2 = 0$ which reduces to $t = 1$, $u = 0$ in virtue of the equation $t^2 + du^2 = 1$. Hence the identity is the only proper automorph taking the solution x_0, y_0 into itself. We have

THEOREM 56. *The number of representations of any number by an indefinite non-zero form is infinite. The number of representations of any number by a positive form or a zero form is the number of essentially distinct solutions multiplied by the number of proper automorphs, that is, by 6, 4 or 2 according as* $d = 3$ *and the form is improperly primitive,* $d = 1$ *or* d *neither* 3 *nor* 1.

Though zero forms are included in the above theory they can be dealt with so simply and completely in the general case that they are worth special consideration. According to theorem 23a any binary zero form is equivalent to $2bxy + cy^2$ where $-b^2 = d$. Then $q = y(2bx + cy)$ is solvable for $y = w$ if and only if w divides q and $q/w \equiv cw \pmod{2b}$ is solvable. Hence we have

THEOREM 57. *If f is a binary zero form of determinant* $d = -b^2$, *the number of solutions of* $f = q$ *is the number of divisors* w *of* q *such that* $q/w \equiv cw \pmod{2b}$.

42. **Ideals in a quadratic field.** Several important results in the theory of binary quadratic forms are obtained by a process called *composition of forms*. This, in turn, is closely connected with the theory of ideals in a quadratic field. Accordingly, in this section we develop some pertinent facts about quadratic ideals.

Let F be a field obtained by adjoining $\sqrt{\Delta}$ to the field of rational numbers, where Δ is a non-square integer congruent to 1 or 0 (mod 4); that is, F consists of all numbers of the form $a + b\sqrt{\Delta}$ where a and b are rational numbers. Let $\sigma = \frac{1}{2}(\sqrt{\Delta} + 1)$ or $\frac{1}{2}\sqrt{\Delta}$ according as $\Delta \equiv 1$ or $0 \pmod 4$ and call *quadratic integers* those numbers expressible in the form $x + y\sigma$ where x and y are rational integers. (We reserve the name "integer" for rational integers.) The set of all quadratic integers in F we denote by $J(\Delta)$ or J. It is easy to see that J is closed under addition, subtraction and multiplication; that is, the sum, difference and product of any two quadratic integers are quadratic integers. If α is an element of J we call the number obtained from α by replacing $\sqrt{\Delta}$ by $-\sqrt{\Delta}$ its *conjugate* and denote the conjugate of α by α^c. Call $\alpha\alpha^c =$

$N(\alpha)$, the *norm* of α, and see that $N(\alpha)N(\beta) = N(\alpha\beta)$. If α is a quadratic integer, $N(\alpha)$ is an integer.

If now $\alpha_1, \alpha_2, \cdots, \alpha_n$ is a set of numbers of J, we call all numbers $\tau_1\alpha_1 + \tau_2\alpha_2 + \cdots + \tau_n\alpha_n$ where the τ_i range over all numbers of J, the *ideal* $I = (\alpha_1, \alpha_2, \cdots, \alpha_n)$. If the α_i are such that every number in I is expressible uniquely in the form $x_1\alpha_1 + x_2\alpha_2 + \cdots + x_n\alpha_n$ where the x_i are integers, we call $\alpha_1, \alpha_2, \cdots, \alpha_n$ a *basis* of I and write $I = [\alpha_1, \alpha_2, \cdots, \alpha_n]$ with brackets instead of parentheses. Two ideals are said to be *equal* if they contain the same numbers. An ideal consisting of all numbers $\alpha\rho$, where α is a fixed number of J and ρ ranges over all numbers of J is called a *principal ideal* and written (α). If $I = [\alpha_1, \alpha_2]$ we call I^c the ideal $[\alpha_1^c, \alpha_2^c]$ and uI the ideal $[u\alpha_1, u\alpha_2]$.

The first classical result we need is

THEOREM 58. *Every ideal I of J has a basis $[r, s + u\sigma]$ where r, s, u are integers, r and u both being positive. Furthermore, r is the* g.c.d. *of the integers in I and u is the* g.c.d. *of the coefficients of σ in the quadratic integers of I. (Hence r and u are uniquely determined by I).*

To prove this first notice that I contains an integer since if α is a quadratic integer in I, then $N(\alpha)$ is also in I. If then we let r be the least positive integer in I we see that it must divide all integers in I since if it does not divide an integer b in I, then $b = qr + r'$ where r' is a positive integer less than r; b and qr being in I implies that r' is in I contrary to the supposition that r has least positive value. By the same reasoning we can show that u has the properties desired.

It remains to show that every number of I is a unique linear combination of r and $s + u\sigma$ with integer coefficients. To this end let $a + b\sigma$ be a number of I. Then $b = ux$ determines an integer x and $a + b\sigma - x(s + u\sigma) = a - x \cdot s$ is an integer in I and hence is equal to ry for some integer y; thus $a + b\sigma = yr + x(s + u\sigma)$. Furthermore $yr + x(s + u\sigma) = y'r + x'(s + u\sigma)$ implies that $x = x'$, $y = y'$ and the representation in terms of the basis is unique. This completes the proof of the theorem.

There are certain restrictions on r, s and u of the theorem above if r, $s + u\sigma$ are to form a basis for an ideal. For instance, 3 and $1 + 2\sqrt{3}$ do not form a basis of an ideal since the equation $(1 - 2\sqrt{3})(1 + 2\sqrt{3}) = -11$ shows that -11 is in the ideal $(3, 1 + 2\sqrt{3})$ while -11 is not expressible in the form $3x + (1 + 2\sqrt{3})y$ for integers x and y. We now find necessary and sufficient conditions on integers r, s and u with $ru \neq 0$ that $(r, s + u\sigma)$ is such a basis. Suppose $(r, s + u\sigma)$ is a basis with $ru \neq 0$. Then all numbers of the ideal are of the form

$$t = r(x_1 + y_1\sigma) + (s + u\sigma)(x_2 + y_2\sigma^c)$$

for integers x_1, x_2, y_1, y_2. The coefficient of σ in t is $ry_1 + ux_2 - sy_2$ and since, from the theory of numbers, y_1, x_2 and y_2 can be chosen so that this coefficient is the g.c.d. of r, u and s, we see that u, being by theorem 58 a divisor of all coefficients of numbers of I, must divide the g.c.d. of r, u and s; hence the equations $r = au$, $s = eu$ determine integers a and e. If, on the other hand, we choose y_1, x_2 and y_2 so that $ry_1 + ux_2 - sy_2 = 0$, that is, $ay_1 + x_2 - ey_2 = 0$, and let $\sigma + \sigma^c = \epsilon$ where ϵ is 1 or 0 according as $\Delta \equiv 1$ or 0 (mod 4) we have

$$(47) \qquad \begin{aligned} t &= rx_1 - (rs/u)y_1 + y_2k/u \\ k &= s^2 + su\epsilon + u^2\sigma\sigma^c = N(s + u\sigma). \end{aligned}$$

Since t is an integer in I it must, by theorem 58, be divisible by r for all values of x_1, y_1 and y_2; hence $k \equiv 0 \pmod{ru}$. Thus we have shown that if $(r, s + u\sigma)$ is a basis under the restrictions of theorem 58, then r and s are divisible by u

and $N(s + u\sigma)$ is divisible by ru. Conversely if these conditions hold we see that u divides all coefficients of σ for numbers of I and that, since all integers in I are expressible in the form of t in (47), r divides all integers in I. Hence we have proved

THEOREM 59. *$I = [r, s + u\sigma]$ for r, s and u integral and $ru \neq 0$ if and only if $r = ua$, $s = ue$ determine integers a and e and $N(s + u\sigma) = 0 \pmod{ru}$. That is, we can write $I = [au, eu + u\sigma] = u[a, e + \sigma]$ with $N(e + \sigma) \equiv 0 \pmod{a}$.*

We call a basis $[r, s + u\sigma]$ of an ideal a *reduced basis* if r, s and u are integers, r and u both being positive. Theorem 59 gives an alternative way of writing a reduced basis.

Suppose $[\alpha_1, \alpha_2]$ and $[\beta_1, \beta_2]$ are two equal or distinct ideals for which the following equations hold:

$$\alpha_1 = \beta_1 t_{11} + \beta_2 t_{21}$$
$$(48) \qquad \alpha_2 = \beta_1 t_{12} + \beta_2 t_{22} .$$

Then we may say that the basis $[\alpha_1, \alpha_2]$ is taken into the basis $[\beta_1, \beta_2]$ by the linear transformation $T = (t_{ij})$. In matrix notation (48) is

$$(\alpha_1, \alpha_2) = (\beta_1, \beta_2)T.$$

Next we prove

THEOREM 60. *Two ideals $[\alpha_1, \alpha_2]$ and $[\beta_1, \beta_2]$ are equal if and only if there is a unimodular transformation taking one basis into the other.*

If there is a unimodular transformation taking one into the other, the numbers in the ideals are the same since (48) and T unimodular implies that $x\alpha_1 + y\alpha_2 = (xt_{11} + yt_{12})\beta_1 + (xt_{21} + yt_{22})\beta_2$ and if the coefficients of α_1 and α_2 are integers, so are the coefficients of β_1 and β_2. Suppose on the other hand, that the ideals are equal. Then there are matrices T and T' with integer elements such that the matric equations $(\alpha_1, \alpha_2) = (\beta_1, \beta_2)T$ and $(\beta_1, \beta_2) = (\alpha_1, \alpha_2)T'$ hold. Then $(\alpha_1, \alpha_2) = (\alpha_1, \alpha_2)T'T$ and since the representation of α_1 and α_2 by α_1 and α_2 is unique we have $T'T = I$, the identity matrix, which shows that $|T'||T| = 1$ and hence $|T| = \pm 1$. Thus the theorem is proved.

If I has the reduced basis $[ua, ue + u\sigma]$ we define the *norm of I*, denoted by $N(I)$, to be ur and see that $N(I) = u^2a$. If I_1 and I_2 are two ideals in J, their product, $I_1I_2 = I_3$, is defined to be that ideal consisting of all numbers in J expressible in the form $\Sigma\rho_{ij}\alpha_i\beta_j$ where the α_i are in I_1, the β_j are in I_2 and the ρ_{ij} are quadratic integers. Since the set of numbers expressible in the form $\Sigma\rho_{ij}\alpha_i^c\beta_j^c$ coincides with the conjugates of the set of numbers expressible in the form $\Sigma\rho_{ij}\alpha_i\beta_j$ we see that $I_1I_2 = I_3$ implies $I_1^cI_2^c = I_3^c$. Multiplication of ideals is associative and commutative since quadratic integers have these properties. We call an ideal $I = [ua, ue + u\sigma]$ *primitive* if 1 is the g.c.d. of a, $2e + \sigma + \sigma^c$ and c where $c = N(e + \sigma)/a$. The following important result holds even if the ideals concerned are neither primitive nor principal but the general proof is much more difficult and the more restricted result is sufficient for our purposes here.

THEOREM 61. *If each of the ideals I_1, I_2 is either primitive in J or principal, then $N(I_1I_2) = N(I_1)N(I_2)$.*

We prove this theorem by showing that it follows from lemma 15 below. This consequence is direct since, using the lemma, $(N(I_1))(N(I_2)) = I_1I_1^cI_2I_2^c = I_3I_3^c = (N(I_3))$, where $I_3 = I_1I_2$, and hence $N(I_1)N(I_2) = N(I_3)$. It remains to show

LEMMA 15. *If the ideal I is primitive or principal, then $I \cdot I^c$ is the principal ideal consisting of all multiplies of $N(I)$ by numbers of J.*

To prove this first for I primitive, write $I = [ua, ue + u\sigma]$ and see that all elements of $I \cdot I^c$ are linear combinations of u^2a^2, $u^2a(e + \sigma)$, $u^2a(e + \sigma^c)$, $u^2aN(e + \sigma)/a = u^2ac$, with integer coefficients. Thus all numbers of $I \cdot I^c$ are divisible by u^2a. Furthermore, addition of the second and third numbers above, shows that $I \cdot I^c$ contains u^2ag where g is the g.c.d. of a, $2e + \sigma + \sigma^c$ and c which, by the primitiveness of I, is equal to 1. This shows that $I \cdot I^c$ is the principal ideal (u^2a).

If, on the other hand, I is principal it may be written in the form $u(v + w\sigma)$ where v and w are relatively prime. Then $I \cdot I^c = (u^2N(v + w\sigma))$ and it remains to show that $N(I) = u^2N(v + w\sigma)$. We do this by finding a reduced basis for $I_1 = (v + w\sigma)$. Now every number of I_1 is expressible in the form $L = (v + w\sigma)(x + y\sigma^c)$ for integers x and y. But $L = vx + wyN(\sigma) + (\sigma + \sigma^c)yv + (wx - yv)\sigma$ shows that x and y may be chosen so that the coefficient of σ is 1. All values of x and y which make $wx - yv = 0$ are integral multiples of $x = v$ and $y = w$. Hence all integers in I_1 are integral multiplies of $v^2 + w^2N(\sigma) + (\sigma + \sigma^c)vw = N(v + w\sigma)$ which is therefore the norm of I_1.

It is convenient to call $d(I)$, the expression $-(\alpha_1^c\alpha_2 - \alpha_2^c\alpha_1)^2/4$, the *determinant of an ideal* $[\alpha_1, \alpha_2]$. To justify this definition by showing that its value is independent of the particular basis chosen let $(\alpha_1', \alpha_2') = (\alpha_1, \alpha_2)T$ where T is unimodular. Then

$$\begin{vmatrix} \alpha_1' & \alpha_2' \\ \alpha_1'^c & \alpha_2'^c \end{vmatrix} = \begin{vmatrix} \alpha_1 & \alpha_2 \\ \alpha_1^c & \alpha_2^c \end{vmatrix} \mid T \mid$$

which implies that the two determinants are equal except perhaps in sign, and hence that the change of basis does not alter $d(I)$. Furthermore, if $I = [r, s + u\sigma]$, where r, s and u have the properties imposed in theorem 59, we have $d(I) = -r^2u^2(\sigma - \sigma^c)^2/4 = -N^2(I)(\sigma - \sigma^c)^2/4 = -N^2(I)\Delta/4$.

43. A correspondence between ideal classes and classes of quadratic forms. Suppose $I = [ua, ue + u\sigma]$. Then $N(uax + u(e + \sigma)y) = u^2N(ax + (e + \sigma)y) = u^2a\,(ax^2 + bxy + cy^2)$ where

$$(49) \qquad b = 2e + \sigma + \sigma^c, \qquad ac = N(e + \sigma).$$

Furthermore $ac - b^2/4 = -\frac{1}{4}(\sigma - \sigma^c)^2 = -\frac{1}{4}\Delta$. Moreover, we have seen that any basis of I may be obtained from its reduced basis by a unimodular transformation. This transformation will leave unaltered the norm and determinant of the ideal and takes f into an equivalent form. Thus we have proved

THEOREM 62. *If $I = [\alpha_1, \alpha_2]$ is in $J(\Delta)$, then*

$$(50) \qquad N(\alpha_1 x + \alpha_2 y) = N(I)f(x, y)$$

where $f(x, y) = ax^2 + bxy + cy^2$ and $d(f) = -\frac{1}{4}\Delta$, that is, $d(I) = N^2(I)d(f)$. If the basis is reduced, the relationship between the basis and the coefficients of f is given by equations (49).

The following theorem establishes the correspondence in the other direction.

THEOREM 63. *With a class of forms there is associated by* (51) *below an ideal I in $J(\Delta)$, where $b^2 - 4ac = \Delta$ and for some basis of I and some form f of the class, equation* (50) *holds with $f = ax^2 + bxy + cy^2$.*

To prove this suppose that m is an integer least in absolute value represented by the class of forms. Then there must be a solution in integers x_0, y_0 of the equation $f = m$. Now x_0 and y_0 must be relatively prime since the square of any common factor would divide m, denying the supposition that m is least. We may then by a unimodular transformation take f into a form whose leading coefficient is m. Hence we write f in the form of theorem 62 with a a non-zero integer least in absolute value of all the numbers represented by f. Then with f we associate the ideal

$$(51)\qquad I = \left(a, \frac{b + \sqrt{b^2 - 4ac}}{2}\right) = (\alpha_1, \alpha_2)$$

in $J(\Delta)$ where $\Delta = b^2 - 4ac$, and see that $N(\alpha_1 x + \alpha_2 y) = af$. It remains to show that $a = N(I)$ and hence (α_1, α_2) is a basis. If $\Delta \equiv 1 \pmod 4$ we may take $r = a$, $s = (b - 1)/2$, $u = 1$ and, since $\sigma = \frac{1}{2}(1 + \sqrt{b^2 - 4ac}) = \frac{1}{2}(1 + \sqrt{\Delta})$, we have $N(s + u\sigma) = ac \equiv 0 \pmod a$ and the conditions of theorem 59 are satisfied. If $\Delta \not\equiv 1 \pmod 4$, b is even and we may take $s = \frac{1}{2}b$, $u = 1$, $\sigma = \frac{1}{2}\sqrt{\Delta}$ and again $N(s + u\sigma) = ac \equiv 0 \pmod a$. Furthermore the correspondence between the coefficients and basis is the same as in (49).

Notice that the form which we have made correspond to I depends on the particular basis but that the class of that form is independent of the basis. In both cases the ideal $I = [\alpha_1, \alpha_2]$ and the form f satisfy condition (50). However, though all forms associated with any ideal by the above means are properly or improperly equivalent to one another, there may be several ideals associated with one form; for instance I and ρI, with ρ a quadratic integer, are associated with the same class of forms. To obtain uniqueness of correspondence we say that two ideals I_1 and I_2 are in the same class (or *ideal class*) if there exist quadratic integers ρ_1 and ρ_2 such that $\rho_1 I_1 = \rho_2 I_2$. Let Σ denote an ideal class (that is, the set of all ideals in a given class) and Γ the set of all forms (properly or improperly) equivalent to a given form. As above, Σ^c denotes the class of ideals obtained from Σ by replacing each number by its conjugate. We prove

THEOREM 64. *Two primitive ideals I_1 and I_2 in $J(\Delta)$ are associated by* (50) *with the same class Γ of forms if and only if the ideals are in the same class Σ or conjugate classes Σ and Σ^c.*

To prove this let $I_i = [\alpha_i, \beta_i]$ and $N(\alpha_i x_i + \beta_i y_i) = N(I_i)f_i(x_i, y_i)$ for $i = 1, 2$ and $f_1(x_1, y_1) \cong f_2(x_2, y_2)$. If T is a unimodular transformation taking f_1 into f_2 we have the equations $(x_1, y_1) = (x_2, y_2)T$ and $f_1(x_1, y_1) = f_2(x_2, y_2)$ identically in x_2 and y_2. Now $(\alpha_1, \beta_1)^T = T^I(\alpha_1', \beta_1')^T$ defines a new basis $[\alpha_1', \beta_1']$ of I_1 and we have $(x_1, y_1)(\alpha_1, \beta_1)^T = (x_2, y_2)(\alpha_1', \beta_1')^T$. Thus $N(\alpha_1 x_1 + \beta_1 y_1) = N(I_1)f_2(x_2, y_2)$ implies $N(\alpha_1' x_2 + \beta_1' y_2) = N(I_1)f_2(x_2, y_2)$ and replacing α_1', β_1' by α_1, β_1 we have

$$N(\alpha_i x_2 + \beta_i y_2) = N(I_i)f_2(x_2, y_2), \qquad i = 1, 2, \tag{52}$$

identically in x_2 and y_2. If a and c are the coefficients of x_2^2 and y_2^2 (they must be different from zero) respectively in $f_2(x_2, y_2)$ we have, taking the pairs of values $(1, 0)$ and $(0, 1)$

$$N(\alpha_i) = N(I_i)a, \qquad N(\beta_i) = N(I_i)c, \tag{53}$$

and hence

$$\begin{aligned} N(\alpha_1)N(I_2) &= N(\alpha_2)N(I_1), \\ N(\beta_1)N(I_2) &= N(\beta_2)N(I_1). \end{aligned} \tag{54}$$

Then the ideals $I_4 = \alpha_1 I_2$ and $I_3 = \alpha_2 I_1$ have equal norms by (54) and theorem 61 and are associated with the same class Γ of forms. Furthermore, I_4 and I_3 are in the same ideal class as I_2 and I_1 respectively. Thus, multiplying (52) by $N(\alpha_2)$ and $N(\alpha_1)$ respectively we have, equating coefficients of x_2^2, x_2y_2, y_2^2 on the left side

$$\begin{aligned} \alpha_4\alpha_4^c = \alpha_3\alpha_3^c, \qquad \alpha_4\beta_4^c + \alpha_4^c\beta_4 &= \alpha_3\beta_3^c + \alpha_3^c\beta_3, \\ \beta_4\beta_4^c &= \beta_3\beta_3^c \end{aligned} \tag{55}$$

where $I_4 = [\alpha_4, \beta_4]$ and $I_3 = [\alpha_3, \beta_3]$. But the determinants of I_4 and I_3 are equal since they depend only on Δ and the norms. Hence

$$\alpha_4\beta_4^c - \alpha_4^c\beta_4 = \pm(\alpha_3\beta_3^c - \alpha_3^c\beta_3). \tag{56}$$

If the positive sign holds we have, adding (56) to the second equation of (55), $\alpha_4\beta_4^c = \alpha_3\beta_3^c$ which, with $\beta_4\beta_4^c = \beta_3\beta_3^c$, implies $\alpha_4/\beta_4 = \alpha_3/\beta_3$ and hence I_4 and I_3 are in the same ideal class. If the negative sign holds, $\alpha_4\beta_4^c = \alpha_3^c\beta_3$ which, with $\beta_4\beta_4^c = \beta_3\beta_3^c$, gives $\alpha_4/\beta_4 = \alpha_3^c/\beta_3^c$ and I_4 and I_3^c are in the same class. This completes the proof that $f_1 \cong f_2$ implies that the corresponding ideal classes are equal or conjugate.

On the other hand, if $\Sigma_1 = \Sigma_2^c$ we may choose $[\alpha_1, \alpha_2]$ and $[\alpha_1^c, \alpha_2^c]$ as bases and $N(\alpha_1 x + \alpha_2 y) = N(\alpha_1^c x + \alpha_2^c y)$ shows that the forms are equal.

Notice that since we call a form primitive when 1 is the g.c.d. of its coefficients, the definition of a primitive ideal and equations (49) show that a form is primitive if and only if the corresponding ideal class is primitive. Then our correspondence will be completed by the following result.

THEOREM 65. *The primitive ideal classes Σ and Σ^c are equal if and only if each corresponding class Γ is improperly equivalent to itself.*

To prove this we may take $[a, e + \sigma]$ as the representative of Σ and $[a, e + \sigma^c]$ as its conjugate. Suppose they are in the same class, that is, there exist quadratic integers α and β such that $\alpha I = \beta I^c$. Then $N(\alpha)I = \alpha^c\beta I^c$ and, since the coefficient of σ in every number of the ideal on the left is divisible by $N(\alpha)$ we see that $N(\alpha)$ divides $\alpha^c\beta$ and that $I = \gamma I^c$ for some quadratic integer γ. Taking the norm of both sides we find that $N(\gamma)$ is 1 and hence that the ideal (γ) is the ideal (1). Thus $I = I^c$ which implies that there exist integers t and v such that $ta + v(e + \sigma) = e + \sigma^c$. Equating coefficients of the rational and irrational parts we have $v = -1$ and $ta = 2e + \sigma + \sigma^c$ which must be solvable for an integer t. Thus from (49) the form corresponding to the ideal is $ax^2 + bxy + cy^2 = f$

in which $b \equiv 0 \pmod{a}$. Then f may be taken into itself by the product of two transformations: first

$$H = \begin{bmatrix} 1 & h \\ 0 & 1 \end{bmatrix}$$

where h is chosen so that $2ah + b$ is 0 or a, and second

$$\begin{bmatrix} 1 & 0 \\ 0 & -1 \end{bmatrix} H^I \quad \text{or} \quad \begin{bmatrix} 1 & 1 \\ 0 & -1 \end{bmatrix} H^I$$

in the respective cases. Thus f is improperly equivalent to itself and the same will be true of every form in the class of f.

We have shown in theorem 52 that if f can be taken into itself by a transformation of determinant -1, there is in the same proper class as f a form f_1 which is taken into itself by the transformation

$$D = \begin{bmatrix} 1 & w \\ 0 & -1 \end{bmatrix}$$

with $w = 1$ or 0. Write $f_1 = ax^2 + bxy + cy^2$ and see that D takes f_1 into $a(x + wy)^2 + b(x + wy)(-y) + cy^2 = ax^2 + xy(2aw - b) + (aw^2 - bw + c)y^2$. Thus $b = 2aw - b$ and $b \equiv 0 \pmod{a}$. Then, retracing our argument in the first part of this theorem we see that the corresponding ideal I has the property that $I^c = I$. This completes the proof and permits us to make the definition: an ideal class Σ is called *ambiguous* if it is equal to its conjugate Σ^c. A class of quadratic forms is *ambiguous* if any (and hence every) form in the class may be taken into itself by a transformation of determinant -1. Hence we have just shown that there is a 1–1 correspondence between ambiguous classes of primitive ideals and ambiguous classes of primitive forms while the correspondence for non-ambiguous classes is 2–2. We therefore understand that in any correspondence between Σ and Γ classes, if Σ corresponds to Γ then Σ^c corresponds to Γ^c. It is this correspondence which enables us to prove expeditiously several important properties of binary quadratic forms.

44. **Composition of ideal classes and classes of forms.** If Γ_1 and Γ_2 are classes of forms associated with the ideal classes Σ_1 and Σ_2 in $J(\Delta)$ we define the class associated with $\Sigma_1\Sigma_2$ in $J(\Delta)$ to be the class $\Gamma_1\Gamma_2$ and this class is said to be derived from Γ_1 and Γ_2 by *composition*. Since Δ determines the determinant of the forms, their determinants are all equal. Furthermore, composition is commutative and associative since the product of ideals has these properties. If $I_i = [\alpha_i, \beta_i]$ we have, from equations (50) that $N(\alpha_i) = N(I_i)a_i$ where a_i is the leading coefficient of the form f_i associated with I_i. If the ideals I_i are primitive, $N(I_1)N(I_2) = N(I_1I_2)$ implies $N(\alpha_1\alpha_2) = N(I_1I_2)a_1a_2$. Thus a_1a_2 will be the leading coefficient of some form in $\Gamma_1\Gamma_2$. Hence we have proved the following important theorem:

THEOREM 66. *If a_1 and a_2 are represented by primitive forms of classes Γ_1 and Γ_2 then a_1a_2 is represented by forms of the class $\Gamma_1\Gamma_2$.*

We need the following two theorems.

THEOREM 67. *A form f is primitive if and only if it represents a number prime to $8d(f)$.*

For f primitive let p_i be any prime factor of $8d(f)$ where $f = ax^2 + bxy + cy^2$. If p_i is prime to a, let $x_i = 1, y_i = 0$; if p_i is a divisor of a and not of c let $x_i = 0, y_i = 1$; if p_i divides both a and c, it does not divide b and we let $x_i = y_i = 1$; for these cases the value of f will be prime to p_i. By the Chinese remainder theorem we can choose

$x \equiv x_i \pmod{p_i}$, $y \equiv y_i \pmod{p_i}$ for all prime divisors p_i of $8d(f)$ and, for such an x and y, f will be prime to $8d(f)$. On the other hand, if g is the g.c.d. of the coefficients of f, it is a divisor of $8d(f)$ and divides all numbers represented by f. Though we have proved this theorem for binary forms a proof along the same lines would establish the same result for any n-ary primitive form.

THEOREM 68. *Two primitive binary forms f and g with integral coefficients are in the same genus if and only if their determinants are equal and there are integers a and w prime to $8d(f)$ such that $f = a$ and $g = w^2a$ are solvable in integers. (This theorem is peculiar to binary forms.)*

We know by theorem 67 that f represents a number a prime to $8d(f)$ and hence we may take f to be $ax^2 + bxy + cy^2$. If f and g are of the same genus we know from the definition of genus that there is a transformation

$$\begin{bmatrix} t_{11} & t_{12} \\ t_{21} & t_{22} \end{bmatrix} w^{-1}$$

taking g into f where the t_{ij} are integers and w is an integer prime to $8d(f)$. Then $(t_{11}/w, t_{21}/w)$ is a solution of $g = a$ and hence $g = aw^2$ is solvable in integers.

Conversely suppose $g = w^2a$ has a solution t_{11}, t_{21} in integers. Then there are integers u and v such that $t_{11}v - t_{21}u = q$ where q is the g.c.d. of t_{11} and t_{21} and hence q^2 divides w^2a. Then the transformation

$$\begin{bmatrix} t_{11}/w & uw/q \\ t_{21}/w & vw/q \end{bmatrix} = T$$

has determinant 1 and takes g into $g' = ax^2 + b'xy + c'y^2$ where b' and c' need not be integers but $qb' \equiv b \pmod 2$. Since q^2 divides w^2a, the denominators of the elements of

T and of b' and c' are prime to $8d(f)$. Then the transformation

$$S = \begin{bmatrix} 1 & (b - b')/2a \\ 0 & 1 \end{bmatrix}$$

takes g' into $g'' = ax^2 + bxy + c''y^2$. Since $d(g'') = d(g) = d(f)$ we have $c'' = c$ and $g'' = f$. Hence TS takes g into f and TS is a transformation with rational elements whose denominators are prime to $8d(f)$. This shows that f and g are in the same genus.

Since the genus of a form is determined by any odd integer prime to the determinant and represented by the form we have

THEOREM 69. *If* $\Gamma_1 \text{ v } \Gamma_1'$ *and* $\Gamma_2 \text{ v } \Gamma_2'$, *then* $\Gamma_1\Gamma_2 \text{ v } \Gamma_1'\Gamma_2'$.

If a form represents 1 we say that it is in the *principal class* and call the corresponding ideal class the *principal ideal class* Σ_0. The genus containing the principal class is called the principal genus. Notice that the principal ideal class contains the ideal $(1) = J$. We now prove

THEOREM 70. *The primitive ideal classes of a given* $J(\Delta)$ *and hence the primitive classes of forms, form a multiplicative group. Also* $\Sigma\Sigma^c = \Sigma_0$ *and* $\Gamma\Gamma^c = \Gamma_0$ *for primitive classes* Σ *and* Γ *and their conjugates.*

First we see that Γ_1 and Γ_2 primitive imply that $\Gamma_1\Gamma_2$ is primitive since theorem 67 shows that forms f_1 and f_2 of Γ_1 and Γ_2 respectively represent numbers a_1 and a_2 prime to $8d(f)$ and hence, by theorem 66, forms of $\Gamma_1\Gamma_2$ represent a number a_1a_2 prime to $8d(f)$ which shows by theorem 67 that $\Gamma_1\Gamma_2$ is primitive.

Now lemma 15 shows that if I is a primitive ideal, $I \cdot I^c$ is the principal ideal $(N(I)) = N(I) \cdot (1)$ and hence $I \cdot I^c$ is in the principal ideal class.

The principal ideal class is the unit element of the group.

We have shown the existence of the inverse and the closure property. Furthermore the associative property holds from the same property for quadratic integers. Hence we have shown that the primitive ideal classes form a group and, if we choose our correspondence so that $\Gamma_0\Gamma = \Gamma$ for every Γ we have the group properties for the classes of primitive forms under composition.

45. **Consequences of the composition of forms.** We first prove the rather startling result embodied in

THEOREM 71. *If h is the number of proper classes of forms in the principal genus for a given determinant, then the number of proper classes in each genus of primitive forms is h.*

Let $\Gamma_0, \Gamma_2, \cdots, \Gamma_h$ be the classes in the principal genus and let Γ be a representative of any other genus. Then $\Gamma\Gamma_0, \Gamma\Gamma_2, \cdots, \Gamma\Gamma_h$ are all in the genus of Γ by theorem 69. No two are in the same class since $\Gamma\Gamma_i = \Gamma\Gamma_j$ would imply, by multiplication by Γ^c, that $\Gamma_i = \Gamma_j$.

Next we have Gauss's celebrated theorem on duplication.

THEOREM 72. *If Γ is any class in the principal genus of primitive forms there is a class Γ_1 such that $\Gamma_1^2 = \Gamma$.*

If g is a form in the principal class it represents 1 and $f \text{ v } g$ implies, by theorem 68, that $f = w^2$ has a solution in integers where w is some integer prime to $8d(f)$. Hence we may consider f to be $w^2x^2 + b'xy + c'y^2$ where b' and c' are integers. Then we may call $I = [w^2, e + \sigma]$ an ideal associated with f.

Our proof will be complete if we can show that $I_1^2 = I$ where $I_1 = [w, e + \sigma]$. Since w is prime to 2Δ, it is also prime to $b' = 2e + \sigma + \sigma^c$ and hence there are integers x and y such that $wx + (2e + \sigma + \sigma^c)y = 1$. Now $I_1^2 = (w^2, w(e + \sigma), (e + \sigma)^2)$ and hence I_1^2 contains $xw(e + \sigma) + y(e^2 + 2e\sigma + \sigma^2) = e + \sigma - yN(e + \sigma)$. But $N(e + \sigma) \equiv 0$

(mod w^2) from theorem 59 and hence $e + \sigma$ is in I_1^2 which shows that $I_1^2 = [w^2, e + \sigma]$ and completes the proof.

Finally we prove

THEOREM 73. *The number of primitive ambiguous classes of given determinant is equal to the number of genera of primitive forms.*

Since class Γ is ambiguous if and only if $\Gamma^2 = \Gamma_0$ we see that the ambiguous classes form a subgroup of the group of classes of all primitive forms of the given determinant. Also $\Gamma_1^2 = \Gamma_2^2$ implies $\Gamma_1 = \Gamma_2\Gamma_a$ where Γ_a is ambiguous. Let q be the number of ambiguous classes, g the number of genera and h the number of classes in each genus. Then gh/q is the number of distinct squares of classes. This, from theorem 72, is equal to h and hence $g = q$.

Notice that if there are q' ambiguous classes in the principal genus, each genus has q' or 0 ambiguous classes.

46. **Genera.** We saw at the end of the previous chapter that each genus has associated with it certain field and ring invariants. In dealing with binary forms these invariants may be expressed in terms of so-called *characters* which we now define.

Given an integer d and a non-zero integer a prime to $2d$, we say that a has the following set of *characters* relative to d for each prime p:

1. For p an odd prime factor of d: $(a \mid p)$.
2. For $p = \infty$: the sign of a when $d > 0$.
3. For $p = 2$:
   a) $a \pmod 4$ if $d \equiv 1 \pmod 4$ or $d \equiv 4 \pmod 8$,
   b) $(-a, -d)_2$ if $d \equiv 2 \pmod 4$,
   c) $a \pmod 8$ if $d \equiv 0 \pmod 8$.

If a is a number prime to $8d$ and represented by a binary form f of determinant d, the above characters of a are called the *characters of the form*. Reference to section 31 shows that two properly primitive or two improperly primitive forms of given determinant are equivalent in $R(p)$ for all primes p dividing $8d$ if and only if they have the same characters. If p does not divide $2d$, the equality of determinants shows that the forms are equivalent in $R(p)$. In $R(\infty)$ either d is negative and both forms have the same index or d is positive and the index is determined by the sign of a. Hence we have

THEOREM 74. *If f and f' are two properly primitive or two improperly primitive forms of determinant d, they are in the same genus if and only if their corresponding characters, listed above, are all equal.*

We are now in a position to determine the number of genera of primitive forms of given determinant d. Observe that theorem 46 shows us that a primitive form having any set of characters exists provided only that they are consistent with the condition

$$(57) \qquad 1 = \prod_p c_p(f) = \prod_p (-a, -d)_p$$

the product being taken over all primes p dividing $8d$ and $p = \infty$ and also consistent with conditions 4 and 5 of theorem 29. Now the choice of a prime to p determines the character of the form; furthermore it affects the value of $(-a, -d)_p$ and hence is restricted by (57) only under the following conditions:

1. The prime p occurs in d to an odd power.
2. If $p = \infty$ with $d > 0$.
3. If $p = 2$, $d = 4^k d_0$, $d_0 \equiv 1 \pmod 4$ with k a non-negative integer.

Condition 4 of theorem 29 automatically holds. Condition 5 becomes

$$i = 1 \text{ if } d < 0, i = 0 \text{ if } d > 0 > a, i = 2 \text{ if } d > 0 < a.$$

We may restrict consideration to finite primes by specifying that if $d > 0$, f shall be positive definite and hence that a is positive. Then if one of these conditions holds, the condition (57) has the effect of dividing by 2 the number of possible sets of characters of a. If none of these conditions occur, that is if $-d$ is a square, any possible set of characters of a is permissible. Notice that for each character listed above there are two possible values except that when $d \equiv 0 \pmod 8$ there are four. Hence we have

THEOREM 75. *Let s be the number of distinct prime factors of d (or $4d$ if $4d$ is an odd integer). The number of distinct genera of primitive forms f of determinant d, f being positive definite if $d > 0$, is*

$$2^{s-1}$$

except that when $d \equiv 1 \pmod 4$ or $d \equiv 0 \pmod 8$, s is increased by 1. If $-d$ is a square the number of genera in the respective cases is doubled.

47. **Reduction of positive definite and zero binary forms.** In chapter III we considered the general reduced form of Hermite. There we showed that every form is equivalent to a reduced form; but two reduced forms might be equivalent. For definite and zero binary forms we can sharpen the inequalities so that the reduced form is unique, that is, so that no two reduced forms are equivalent.

As in the previous section we write $f = ax^2 + bxy + cy^2$, a, b, c being integers, and $d = ac - b^2/4$. When the form is definite, we shall, without loss of generality, take it to be positive definite.

For f positive definite we now prove

THEOREM 76. *Any positive definite binary form with integral coefficients of determinant d is properly equivalent* *to a reduced form* $f = ax^2 + bxy + cy^2$ *whose coefficients satisfy the following inequalities:*

$$(58) \qquad -a < b \leq a \leq c; \quad \text{if } a = c, \quad \text{then } b \geq 0.$$

It follows that $a \leq 2\sqrt{d/3}$, *no two reduced forms are properly equivalent, and* a *is the least positive integer represented by the form.*

Since the reduced form of this theorem is reduced according to theorem 23, $a \leq 2\sqrt{d/3}$ from that theorem. First we show that every form is properly equivalent to a form satisfying conditions (58). By replacing x by $-y$ and y by x in f, if necessary, we may make $a \leq c$. The transformation

$$Q = \begin{bmatrix} 1 & u \\ 0 & 1 \end{bmatrix}$$

takes f into a form whose middle coefficient is $2au + b = b'$ which, by proper integral choice of u may be made not greater in absolute value than a. By consecutive use of these two transformations we can make $|b| \leq a \leq c$. If $a = -b$, the transformation Q with $u = 1$ takes $ax^2 - axy + cy^2$ into $ax^2 + axy + cy^2$. If $a = c$ and $b < 0$, replace x by $-y$, y by x, which changes the sign of the middle coefficient and leaves unaltered the other two.

Next we show that if conditions (58) hold and

$$(59) \quad as^2 + bst + ct^2 = e \leq a, \text{ with } s \text{ and } t \text{ integral}$$

then one of the following holds:

1. $t = 0$, $s^2 = 1$, $a = e$,
2. $t = \pm 1$, $a = e = c$, $s = 0$,
3. $t = \pm 1$, $a = e = c$, $s^2 = 1$, $a = \mp bs$.

Now (59) implies

$$(as + \tfrac{1}{2}bt)^2 + dt^2 = ae \leqq 4d/3.$$

If $t = 0$, this implies $as^2 \leq a$ and hence $s^2 = 1$. Otherwise $t = \pm 1$ and $e = as^2 \pm bs + c \leq a$ with $a \geq | b |$ implies condition 2 or 3. This also shows that a is the least positive integer represented by the form.

To show that two equivalent forms f and f' whose coefficients satisfy (58) must be identical notice first that their leading coefficients must be equal since they represent the same numbers. Let

$$T = \begin{bmatrix} s & u \\ t & v \end{bmatrix}$$

be a properly unimodular transformation taking f into f'. Then $as^2 + bst + ct^2 = a$ and, by the previous paragraph we have one of the three conditions holding. If $t = 0$, $s = v = \pm 1$, then $b' = \pm 2au + b$ which implies $| b' - b | = 2a | u |$. But (58) gives $u = 0$ or $u^2 = 1$. In the former case $b = b'$ and in the latter case $| b | = | b' | = a$ when (58) shows that b and b' are positive and hence equal. But $| f | = | f' |$ and $b = b'$ implies $c = c'$ and the forms are identical. On the other hand if $t = \pm 1$ and $s^2 = 1$ we may multiply T by the automorph $-I$ if necessary to make $s = 1$, $t = \pm 1$, $a = \mp b$ which, from (58), implies $a = b$, $t = -1$. Hence $u + v = -1$ and $b' = a(v - u) = a(-1 - 2u)$ which implies $u = -1$, $b' = a$ and $f = f'$. Condition 2 is similarly disposed of.

COROLLARY 76. *There is a finite number of classes of integral binary forms of given determinant.*

This corollary follows from theorem 76 since a is bounded by d and b by a. As a matter of fact, the theorem can easily be shown if the coefficients are rational and the corollary if the coefficients are rational with bounded denominators.

For zero forms we have the following

THEOREM 77. *If f is a zero form of determinant $d = -r^2$, where r is a positive integer, it is properly equivalent to a reduced form $f_0 = 2rxy + ey^2$ with $| e | < r$ or $e = r$. No two reduced forms are equivalent. See theorem 23a.*

First see that since f represents 0 with the variables not both zero it is equivalent to a form $2rxy + ey^2$ where $r^2 = -d$. If $r < 0$, the coefficient of xy may be made positive by use of a properly unimodular transformation whose first column is e/g, $-2r/g$ with $g = (e, 2r)$. Then, replacing x by $x + uy$ takes $2rxy + ey^2$ into $2rxy + (e + 2ru)y^2$ and an integral u may be chosen so that $| e + 2ru | < r$ or $e + 2ru = r$. This shows that every zero form is equivalent to a reduced form.

Now suppose f_0 and $f_0' = 2rxy + e'y^2$ are equivalent and $| e' | < r$ or $e' = r$. If

$$T = \begin{bmatrix} s & u \\ t & v \end{bmatrix}$$

takes f_0 into f_0' we have $t(2rs + et) = 0$. If $t = 0$ we can, along the lines of the proof of theorem 76 show that $e = e'$. If $t \neq 0$, the first column of T is e/g, $-2r/g$ which, by the previous paragraph is seen to change the sign of r.

COROLLARY 77. *The number of classes of integral zero forms of determinant $-r^2$ is $2r$.*

48. **Reduction of indefinite, non-zero, binary forms.** Let the form f and its determinant d be determined as in the previous section with $-d$ a non-square. We have seen from theorem 23 that every non-zero form is equivalent

to a form whose coefficients satisfy the following inequalities:

$$|b| \leq |a| \leq |c|. \tag{60}$$

Notice that this with d negative implies that $-ac > 0$.

We shall call such a form *semi-reduced*. Even if we used the sharper inequalities of the previous section a form would not necessarily be unique in its class. For example $f = x^2 - 23y^2$ and $g = 2x^2 + 2xy - 11y^2$ satisfy the sharper inequalities but the transformation

$$\begin{bmatrix} 2 & 9 \\ 1 & 5 \end{bmatrix}$$

takes g into f. This section is devoted to the development of a method of finding whether or not two semi-reduced indefinite non-zero binary forms are equivalent.

Our methods depend on the use of continued fractions defined as follows. If

$$\theta = a_1 + \cfrac{1}{a_2 + \cfrac{1}{a_3 + \cdots}}$$

then θ is said to be expanded into a *continued fraction* and is written more compactly as

$$\theta = \{a_1, a_2, a_3, \cdots\}$$

where the a_i may be finite or infinite in number. If the a_i are all positive integers except for the first which may also be zero it is called a *simple continued fraction*. Since this is the only type we shall consider here, we omit the adjective "simple". The first k terms of any expansion constitute a rational number p_k/q_k in lowest terms which is called the *k-th convergent* of the continued fraction. Two results from elementary continued fraction theory we now state without proof.

LEMMA 16. *If* $| \theta - p/q | < 1/2q^2$ *then* p/q *is a convergent in the expansion of* θ *as a continued fraction,* p *and* q *being positive integers.*

LEMMA 17. *If* θ *is a root of a quadratic equation with integral coefficients, its continued fraction expansion is periodic, that is, there are integers* k *and* r *and a real number* $\varphi > 1$ *(*φ *is actually a root of a quadratic equation with integral coefficients) such that*

$$\theta = \{a_1, a_2, \cdots, a_k, \varphi\} = \{a_1, a_2, \cdots, a_{k+r}, \varphi\}$$

r being the length of the period.

Suppose f and f' are two semi-reduced equivalent forms. Now, the equivalence of any of the following pairs implies the equivalence of any other:

$$\begin{array}{ll} ax^2 + bxy + cy^2, & a'x^2 + b'xy + c'y^2 \\ ax^2 - bxy + cy^2, & a'x^2 - b'xy + c'y^2 \\ -ax^2 - bxy - cy^2, & -a'x^2 - b'xy - c'y^2. \end{array}$$

Hence we may assume without loss of generality

$$0 \le b \le a, \qquad a \ge | a' |. \tag{61}$$

Suppose there is a transformation whose first column is (p, q) with $p > 0 < q$ and which takes f into f'. Then

$$ap^2 + bpq + cq^2 = a' \tag{62}$$

and hence

$$(p/q - \theta)(p/q + \theta') = a'/aq^2 < 1 \text{ in absolute value,}$$

where $\theta = (\sqrt{-d} - b/2)/a$ and $\theta' = (\sqrt{-d} + b/2)/a$.

First if $a' > 0$, then $p/q > \theta > 0$. Hence $p/q + \theta' >$

$2\sqrt{-d}/a \geq 2$ and from lemma 16 p/q is a convergent in the expansion of the quadratic surd θ.

Second if $a' < 0$, we obtain from (62)

$$(q/p - 1/\theta)(q/p + 1/\theta') = a'/cp^2 > 0.$$

Hence $q/p > \theta^{-1}$ and $q/p + 1/\theta' > 2\sqrt{-d}/-c$. Then $(2\sqrt{-d}/-c)(c/a') \geq 2$ shows that q/p is a convergent of θ^{-1}. Hence p/q is a convergent of θ as may be seen by comparing the expansions of θ and θ^{-1} as follows: If $\theta = \{a_1, a_2, \cdots\}$, $a_1 > 0$, then $\theta^{-1} = \{0, a_1, a_2, \cdots\}$. If $a_1 = 0$, reverse the argument. Thus we have proved

THEOREM 78. *If p, q is the first column of a properly unimodular transformation taking f into $f' = a'x^2 + b'xy + c'y^2$ with p and q positive and*

$$0 \leq b \leq a \geq | a' |$$

and f is semi-reduced as well as f', then p/q is a convergent of the continued fraction expansion of

$$\theta = (\sqrt{-d} - b/2)/a.$$

To complete our theory we shall prove two theorems:

THEOREM 79. *If $f \cong f'$, there is a unimodular transformation whose first column consists of positive integers and which takes f into f'.*

THEOREM 80. *If θ defined in theorem 78 is expressible in the forms of lemma 17, where $p/q = \{a_1, a_2, \cdots, a_k\}$ and $p'/q' = \{a_1, a_2, \cdots, a_{k+r}\}$ then there is an automorph P of f such that*

$$\begin{bmatrix} p' \\ q' \end{bmatrix} = P \begin{bmatrix} p \\ q \end{bmatrix}.$$

Both of these theorems are proved using results on the

automorphs of the binary form. We know from theorem 50 that

$$P = \begin{bmatrix} t - \frac{1}{2}bu & -cu \\ au & t + \frac{1}{2}bu \end{bmatrix}$$

is an automorph of f for t, u any semi-integral solution of $t^2 + du^2 = 1$. Let

$$T = \begin{bmatrix} p & r \\ q & s \end{bmatrix}$$

be a properly unimodular transformation taking f into f'. Then $T' = PT$ takes f into f' and, letting $p'q'$ be the first column of T' we have

$$p' = (t - \tfrac{1}{2}bu)p - cuq,$$

$$q' = aup + (t + \tfrac{1}{2}bu)q$$

and $p'q' = uap^2(t - \frac{1}{2}bu) + pq - cuq^2(t + \frac{1}{2}bu)$. Now $t^2 + du^2 = 1$ implies $(t - \frac{1}{2}bu)(t + \frac{1}{2}bu) = 1 - acu^2 > 0$. Hence if we make t positive one of the terms of the product is positive and hence the other. Thus choose the sign of u so that ua is positive which implies $-cu$ is positive. Then u and t may be chosen so large that $p'\,q'$ is positive. If p' and q' are both negative replace T' by $-T'$ and prove the theorem.

To prove theorem 80, we need the following

LEMMA 18. *If* $\theta = \{a_1, a_2, \cdots, a_k, \varphi\}$, *then*

$$\theta = \frac{p\varphi + r}{q\varphi + s}$$

where

$$\begin{bmatrix} p & r \\ q & s \end{bmatrix} = \begin{bmatrix} a_1 & 1 \\ 1 & 0 \end{bmatrix}\begin{bmatrix} a_2 & 1 \\ 1 & 0 \end{bmatrix} \cdots \begin{bmatrix} a_k & 1 \\ 1 & 0 \end{bmatrix}$$

and p/q, r/s are the k-th *and* $(k - 1)$st *convergents in the expansion of* θ.

To prove this by induction first notice that it holds for $k = 1$. Let $a_k + \varphi^{-1} = \varphi_1$. Then, by the hypothesis of our induction

$$\theta = \frac{p_1 \varphi_1 + r_1}{q_1 \varphi_1 + s_1}$$

where

$$\begin{bmatrix} p_1 & r_1 \\ q_1 & s_1 \end{bmatrix} \begin{bmatrix} a_k & 1 \\ 1 & 0 \end{bmatrix} = \begin{bmatrix} p & r \\ q & s \end{bmatrix}.$$

Then, substituting the resulting values of p_1, r_1, q_1, s_1 and φ_1, we get the desired expression for θ. The rest of the lemma is proved by letting φ become infinite and approach zero.

Then, returning to the proof of the theorem, we have

$$\theta = \frac{p\varphi + r}{q\varphi + s} = \frac{p'\varphi + r'}{q'\varphi + s'}.$$

Hence, for properly chosen constants ρ and ρ'

$$\begin{bmatrix} \theta \\ 1 \end{bmatrix} = \rho \begin{bmatrix} p & r \\ q & s \end{bmatrix} \begin{bmatrix} \varphi \\ 1 \end{bmatrix} = \rho' \begin{bmatrix} p' & r' \\ q' & s' \end{bmatrix} \begin{bmatrix} \varphi \\ 1 \end{bmatrix}$$

which shows that

$$\begin{bmatrix} p' & r' \\ q' & s' \end{bmatrix} = P \begin{bmatrix} p & r \\ q & s \end{bmatrix}$$

for some properly unimodular transformation P having the property that

$$P \begin{bmatrix} \theta \\ 1 \end{bmatrix} = \rho'' \begin{bmatrix} \theta \\ 1 \end{bmatrix}.$$

Hence, if $P = (p_{ij})$ we have

$$\theta = \frac{p_{11}\theta + p_{12}}{p_{21}\theta + p_{22}}$$

which implies $p_{21}\theta^2 + (p_{22} - p_{11})\theta - p_{12} = 0$. But $a\theta^2 + b\theta + c = 0$, which, since f is primitive, implies for some integer u,

$$p_{21} = ua, \qquad p_{22} - p_{11} = ub, \qquad -p_{12} = uc.$$

Let $t = \frac{1}{2}bu + p_{11}$ and see that $p_{22} = t + \frac{1}{2}bu$. Hence $1 = p_{22}p_{11} - p_{21}p_{12} = t^2 + du^2$, which shows that P is an automorph of f.

Thus we have shown

THEOREM 81. *If there is a transformation taking f into f' there is one whose first column appears as a convergent in the portion of the continued fraction expansion of θ, as defined in theorem* 78, *through the repeating part, where f and f' are semi-reduced and satisfy* (61).

Of course the first column does not determine the transformation but if T_1 and T_2 are two unimodular transformations whose first columns are the same, the following relationship exists between them:

$$T_1 = T_2\begin{bmatrix} 1 & t \\ 0 & 1 \end{bmatrix}$$

for an integer t, as may be seen by computing $T_2^I T_1$. The value of t is determined with a choice of at most two values by the requirement that the forms be semi-reduced,

We illustrate the process by considering $f = 2x^2 + 2xy - 11y^2$ and finding all semi-reduced forms equivalent to it whose leading coefficient is not more than 2 in absolute value. Here $d = -23$ and

$$\theta = (-1 + \sqrt{23})/2.$$

The expansion of θ as a continued fraction is

$$\{1, \overline{1, 8, 1, 3,} 1, 8, 1, 3, \cdots\}$$

where the line is drawn over the repeating part. Hence the possible first columns of transformations taking f into equivalent semi-reduced forms are obtained from the first five convergents. Writing $f = f(x, y)$ and computing the convergents with the aid of lemma 18, we have

Convergent	
1/1	$f(1, 1) = -7$
2/1	$f(2, 1) = 1$
17/9	$f(17, 9) = -7$
19/10	$f(19, 10) = 2$
74/39	$f(74, 39) = -7$

All but the second and fourth are excluded since the value of the form is greater in absolute value than 2. For 2/1, use the transformation

$$\begin{bmatrix} 2 & 1 \\ 1 & 1 \end{bmatrix}$$

which takes f into $x^2 - 8xy - 7y^2$ which is equivalent to $x^2 - 23y^2$. For 19/10 use

$$\begin{bmatrix} 19 & -2 \\ 10 & -1 \end{bmatrix}$$

which takes f into $g = 2x^2 - 10xy + y^2$. Then

$$\begin{bmatrix} 1 & t \\ 0 & 1 \end{bmatrix}$$

will take g into a semi-reduced form for $t = 2, 3$. Thus the only semi-reduced forms equivalent to f and whose leading coefficients are less than 3 in absolute value are: f, $x^2 - 23y^2$, $2x^2 - 2xy - 11y^2$.

49. **Class number.** One of the most important functions connected with binary forms is the *class number*, that is, the number of proper classes of forms of given determinant. We showed in section 45 that for primitive forms this is a multiple of the number of genera. The class number of forms of determinant d can be expressed as a multiple of the Dirichlet series

$$\Sigma(-d \mid n)n^{-1}$$

the sum being over all positive integers n prime to $2d$. It can also be expressed in finite terms. We have not space for this development here but the reader is referred to a self-contained account in Mathews' "Theory of Numbers", chapter VIII.

However, there is an interesting relationship discovered by Dirichlet, between the class numbers of improperly primitive forms and properly primitive forms of given determinant. Since we shall find this of use in the next chapter, we derive this relationship.

Let

$$f_1 = a_1x^2 + 2b_1xy + c_1y^2, \qquad f_2 = 2a_2x^2 + 2b_2xy + 2c_2y^2$$

where a_1 or c_1 is odd, $a_1c_1 - b_1^2 = 4a_2c_2 - b_2^2 = d \equiv 3 \pmod 4$. Since f represents an odd integer we may assume a_1 to be odd and by replacing x by $x + y$ if necessary assume c_1 is even. Similarly a_2 may be taken to be odd. Thus we assume

$$(63) \qquad a_1 \text{ odd}, \qquad c_1 \text{ even}, \qquad a_2 \text{ odd}.$$

This implies that b_1 and b_2 are odd and $c_1 \equiv 0 \pmod 4$. Take

$$T = \begin{bmatrix} 2 & 0 \\ 0 & 1 \end{bmatrix}$$

and see that T takes $f_1/2$ into $2a_1x^2 + 2b_1xy + \frac{1}{2}c_1y^2$ which is a form f_2 having the properties (63) and determinant d. On the other hand T^I takes $2f_2$ into $a_2x^2 + 2b_2xy + 4c_2y^2$ which is a form f_1 having properties (63) and determinant d.

Thus we have a 1–1 correspondence between forms f_1 and f_2. We wish to show that under certain circumstances this establishes a 1–1 correspondence between classes and to determine the correspondence in all circumstances. Suppose

$$S_1 = \begin{bmatrix} p_1 & q_1 \\ r_1 & s_1 \end{bmatrix}$$

takes f_1 into f_1', S_1 being properly unimodular. If T takes $f_1'/2$ into f_2', then $T^I S_1 T$ takes f_2 into f_2'. We need to show that the latter transformation has integral elements and hence $f_2 \cong f_2'$. Now f_1 and f_1' having properties (63) imply q_1 is even,

$$T^I S_1 T = \begin{bmatrix} p_1 & \frac{1}{2}q_1 \\ 2r_1 & s_1 \end{bmatrix}$$

and our equivalence is shown.

Suppose

$$S_2 = \begin{bmatrix} p_2 & q_2 \\ r_2 & s_2 \end{bmatrix}$$

takes f_2 into f_2', S_2 being properly unimodular. If T^I takes f_2' into f_1', then TS_2T^I takes f_1 into f_1'. But

$$TS_2T^I = \begin{bmatrix} p_2 & 2q_2 \\ \frac{1}{2}r_2 & s_2 \end{bmatrix}.$$

This will have integer elements if and only if r_2 is even. Since S_2 takes f_2 into another form having property (63) we have

$$a_2p_2^2 + b_2p_2r_2 + c_2r_2^2 \equiv 1 \pmod 2.$$

Thus $(2a_2p_2 + b_2r_2)^2 + dr_2^2 \equiv 4 \pmod 8$ and

$$(64) \qquad r_2 \text{ is even if } d \equiv 7 \pmod 8.$$

Furthermore, there will be a properly unimodular transformation taking f_1 into f_1' if there is an automorph P of f_2 such that, in the matrix PS_2, the element in the lower left corner is even. Theorem 50 gives us

$$PS_2 = \begin{bmatrix} t - \frac{1}{2}b_2u & -c_2u \\ a_2u & t + \frac{1}{2}b_2u \end{bmatrix} \begin{bmatrix} p_2 & q_2 \\ r_2 & s_2 \end{bmatrix}$$

where t, u is a semi-integral solution of $t^2 + (d/4)u^2 = 1$. Then we wish to make $r_2' = a_2p_2u + (t + \frac{1}{2}b_2u) \cdot r_2$ even. If t is half an odd number we may, by changing the sign of u, make $t + \frac{1}{2}b_2u$ odd or even at pleasure and hence r_2' even. If t is an integer, this cannot be done. Hence we have a 1–1 correspondence between classes if $d \equiv 7 \pmod 8$ or if

(65) there is a solution of $x^2 + dy^2 = 4$ with x odd.

If then d is positive we see that (65) holds only if $d = 3$. Hence in what follows we consider only

(66) $d \equiv 3 \pmod 8$ and either
1) $d > 3$, or
2) $d < 0$ and all solutions of $x^2 + dy^2 = 4$ have x even.

Now $d \equiv 3 \pmod 8$ implies that c_2 is odd. We shall show that under this condition, there is a 3–1 correspondence between the classes of forms f_1 and the classes of forms f_2. Let

$$R_1 = \begin{bmatrix} 1 & 0 \\ 0 & 1 \end{bmatrix}, \quad R_2 = \begin{bmatrix} 0 & 1 \\ -1 & 0 \end{bmatrix}, \quad R_3 = \begin{bmatrix} 0 & 1 \\ -1 & 1 \end{bmatrix},$$

$$TR_i = T_i, \; i = 1, 2, 3.$$

If T_i takes f_1 into $2f_{2i}$ then $T_i^I T_j = R_i^I R_j$ takes f_{2i} into f_{2j} and R_i being unimodular shows that $f_{2i} \cong f_{2j}$; f_{2i} may be seen to have property (63) for $i = 1, 2, 3$.

On the other hand, suppose T_i^I takes $2f_2$ into f_{1i}. Then f_{1i} satisfy conditions (63) and we wish to show $f_{1i} \not\cong f_{1j}$ for $i \neq j$. If a properly unimodular transformation R took f_{1i} into f_{1j}, then $T_i^I R T_j$ would take f_2 into itself. Since $T_i^I R T_j$ takes f_2 into itself if and only if $(T_i^I R T_j)^I$ does, we have only $i < j$ to consider. We have to show that such automorphs cannot exist without denying conditions (66).

$$T_1^I R T_2 = \begin{bmatrix} -\tfrac{1}{2} r_{12} & r_{11} \\ -r_{22} & 2r_{21} \end{bmatrix}$$

where $R = (r_{ij})$. The fact that it is an automorph of f^2 implies that r_{12} is even, hence r_{22} odd and, using theorem 50, $r_{22} = -a_2 u$ implies that u is odd contrary to condition (66). Next

$$T_2^I R T_3 = \begin{bmatrix} r_{22} & -2r_{21} - r_{22} \\ -\tfrac{1}{2} r_{12} & r_{11} + \tfrac{1}{2} r_{12} \end{bmatrix}$$

and as above, r_{12} is even. Hence r_{11} and r_{22} are odd. But r_{22} odd and $-2r_{21} - r_{22} = -c_2 u$ implies u is odd. Finally

$$T_1^I R T_3 = \begin{bmatrix} -\tfrac{1}{2} r_{12} & r_{11} + \tfrac{1}{2} r_{12} \\ -r_{22} & 2r_{21} + r_{22} \end{bmatrix}$$

which is dealt with as in the first case.

We summarize our results in the following

THEOREM 82. *Let $h(d)$ be the number of classes of properly primitive integral forms of determinant d with* $d \equiv 3 \pmod 4$ *and $h'(d)$ the number of classes of improperly primitive forms of the same determinant. Then $h(d) = h'(d)$ except when* $d \equiv 3 \pmod 8$ *and one of the following conditions hold:*

1) $d > 3$,
2) $d < 0$ *and all integral solutions of* $x^2 + dy^2 = 4$ *have x even.*

In the exceptional cases, $h(d) = 3h'(d)$.

CHAPTER VIII

TERNARY QUADRATIC FORMS

50. Numbers represented by ternary genera. In this chapter we shall consider only forms whose matrices have integral elements. We have seen in theorem 40 that the genus of a form is characterized by the forms to which it is congruent for an arbitrary modulus. In the previous chapter we showed that the genus of a primitive binary form could be characterized by the quadratic character of numbers represented by the form, where the characters were taken with respect to the prime factors of twice the determinant. For ternary forms (but not for forms in more than three variables) the genus may also be characterized by the numbers represented by the forms but in a slightly different sense.

From theorem 36 and corollary 14, any ternary form f represents any number N in $R(p)$ for any prime p not dividing $2d$ with $d = |f|$. Hence $f \equiv N \pmod{p^k}$ is solvable for k arbitrary if p does not divide $2d$. Thus from corollary 44a we know that N will be represented by some form in the genus of f if it is so represented in the field of reals and if

$$f \equiv N \pmod{p^{r+1}} \tag{67}$$

is solvable for every prime p dividing $2d$, where p^r is the highest power of p dividing N or $4N$ according as p is odd or even. Conversely if (67) is not solvable for some such p, no form in the genus of f represents any number in the arithmetic progression

$$np^{r+1} + N, \qquad n = 0, \pm 1, \pm 2, \cdots .$$

Thus there will be a finite number of arithmetic progressions of this type which have the following properties:

1. No number in any such progression is represented by a form in the genus of f.
2. Every number not in any of these progressions and of the proper sign if f is definite, will be represented by a form in the genus of f.

Such a set of progressions then characterize any genus of ternary forms. We give two examples. See also the discussion at the close of section 33.

EXAMPLE 1. Let $f = x^2 + y^2 + z^2$. The only progressions of the genus involve powers of 2. If N is odd, $f \equiv N \pmod 8$ is solvable except for $N \equiv 7 \pmod 8$; if $N \equiv 2 \pmod 4$, $f \equiv N \pmod{16}$ is solvable; if $N \equiv 0 \pmod 4$, $f \equiv N \pmod 4$ implies x, y, z are even and f represents N if and only if it represents $N/4$. Hence a form in the genus of f will represent N if and only if $N \neq 4^k(8n + 7)$. Since, from corollary 23, all forms of determinant 1 are equivalent to f, it represents exclusively those positive integers not of the form $4^k(8n + 7)$.

EXAMPLE 2. Let $f = x^2 + y^2 + 10z^2$, $g = 2x^2 + 2y^2 + 3z^2 - 2xz$. It may be shown (see section 51) that every form in the genus of f is equivalent to f or to g. As in example 1 we can show that the progressions excluded are

$$4^k(16n + 6),$$

Hence all positive numbers not in either of these progressions are represented by f or g.

51. Numbers represented by forms. If a genus of forms contains only one class we know definitely that every number not excluded for congruential reasons, that is, not lying in certain arithmetic progressions is represented by the form. For positive forms, the number of such genera is apparently limited. At any rate there are less than 100 such primitive classes[8] of the form $ax^2 + by^2 + cz^2$.

Of numbers represented by forms in genera of more than one class there is little known. In some cases it has been shown that every number represented by forms of the genus is represented by a particular form. For instance, in the genus of $f = x^2 + 3y^2 + 36z^2$ there are two classes represented by f and $g = 3x^2 + 4y^2 + 9z^2$. It has been shown[8] that f and g represent the same numbers except that g represents primitively no m^2 with $m \equiv 1 \pmod 3$.

From these results it may be seen that it is of importance to determine whether or not two given forms are equivalent. For positive forms there are several definitions of a reduced form which insure uniqueness. The most used is that of Eisenstein which defines

$$f = ax^2 + by^2 + cz^2 + 2ryz + 2sxz + 2txy$$

to be reduced under the following conditions:

1. r, s, t are all positive or all non-negative.
2. $a \leq b \leq c$, $a + b + 2r + 2s + 2t \geq 0$.
3. $a \geq |2s|$, $a \geq |2t|$, $b \geq |2r|$.
4. If $a = b$, $|r| \leq |s|$; if $b = c$, $|s| \leq |t|$; if $a + b + 2r + 2s + 2t = 0$, $a + 2s + t \leq 0$.
5. For $r, s, t \leq 0$: if $a = -2t$, $s = 0$; if $a = -2s$, $t = 0$; if $b = -2r$, $t = 0$.
6. For $r, s, t > 0$: if $a = 2t$, $s \leqq 2r$; if $a = 2s$, $t \leq 2r$; if $b = 2r$, $t \leq 2s$.

The proof of this result is very tedious but its application is simple. One can quickly find the reduced form equivalent to any given one and to find whether or not two forms are equivalent one need merely compare their reduced forms. There are tables of Eisenstein reduced forms of determinant less than 200^5. In such a table it may be seen for instance, that the only reduced forms of determinant 10 are

$$x^2 + y^2 + 10z^2, \qquad 2x^2 + 2y^2 + 3z^2 - 2xz,$$
$$x^2 + 2y^2 + 5z^2.$$

For indefinite forms there seems to be no practicable method of defining a unique reduced form. This is hardly surprising since such a situation exists even for binary forms.

Asymptotic results for ternary forms are very obscure.

There is a remarkable result of Meyer which we must mention even though its proof is too long to be included here. Let Ω be the g.c.d. of the two rowed minors of A, the matrix of f. Now, by lemma 4, there are unimodular matrices P and Q such that PAQ is a diagonal matrix. By lemma 5, Ω is the g.c.d. of the 2 rowed minors of PAQ and hence divides two diagonal elements of PAQ; thus Ω^2 divides $|PAQ| = |A|$ and we can write $d = \Omega^2\lambda$ where λ is an integer. We call the integral form whose matrix is $A^I d/\Omega$ the *reciprocal form* of f. Meyer's theorem is

THEOREM 83. *A genus of indefinite ternary forms f has but one class if the following conditions hold:*

1. *The form f and its reciprocal form are both properly primitive or f is improperly primitive.*
2. *The numbers Ω and λ are relatively prime and neither is divisible by* 4.

It is not known whether or not every genus of indefinite forms has but one class.

Making use of previous theory we can easily investigate zero and universal ternary forms, keeping in mind that here we are concerned with integral representations and forms whose matrices are integral. From theorem 19, any ternary form f which is a zero form in $F(p)$ for p finite or

$p = \infty$ is a universal form in $F(p)$ and conversely (as a matter of fact, this may be also said of any n-ary forms if $n \neq 4$). Hence if f is universal in the ring of integers, it is universal in $F(p)$ for all p, hence is a zero form in the field of rationals from theorem 27, and thus is a zero form in the ring of rational integers. This proves

THEOREM 84. *Any universal ternary form (in fact every universal form in n variables, $n \neq 4$) is a zero form.*

The converse of this theorem is not true as shown by the fact that $x^2 + y^2 - 49z^2$, which is a zero form, does not represent 7.

52. The number of representations by ternary genera. If we specialize to ternary forms our general results of section 38 we obtain an interesting connection with the class number function. We confine ourselves in this section to the representations of an integer q prime to twice the determinant of the form. (We could similarly apply the methods of section 38 to find the number of representations of binary forms by ternary forms.) From theorem 49b we have

$$G(A, q) = 2^{\mu(q)} \Sigma \lambda_E^{-1}$$

where A is the matrix of f, $G(A, q)$ is the number of essentially distinct primitive representations of q by the classes of the genus of f, $\mu(q)$ is the number of distinct prime factors of q increased by 1 if $q \equiv 0 \pmod 8$ and decreased by 1 if $q \equiv 2 \pmod 4$; neither of these special cases occur for the q considered in this section. Also λ_E has to do with the automorphs of forms E of determinant dq and d is the determinant of f. Furthermore, all E belong to one or at most two genera and if one class of the genus of E is included all classes in the genus also occur.

We first characterize more definitely the forms E which

can occur for any genus of forms of matrix A. Recall that with every representation of q by A is associated a transformation taking A into

$$A_0 = \begin{bmatrix} q & C \\ C^T & D \end{bmatrix}$$

and that E is the 2 by 2 matrix $qD - C^TC$. Let Ω be the g.c.d. of the two-rowed minor determinants of A. We showed before theorem 83 that Ω^2 divides d and hence is prime to q. Since the transformation

$$\begin{bmatrix} 1 & -q^{-1}C \\ 0 & 1 \end{bmatrix}$$

has rational elements with denominators prime to $2d$ and takes A_0 into $q \dotplus Eq^{-1}$, we see that Ω is the g.c.d. of the elements of E. Hence $E/\Omega = E'$ is the matrix of a properly or improperly primitive form $g = ax^2 + 2bxy + cy^2$ whose determinant is $\Delta = qd/\Omega^2$. We shall prove

THEOREM 85. *The following characterize the classes g.*

1. *If* $\Delta \equiv 0$, 1 or 2 (mod 4) *all classes g are properly primitive.*
2. *If* $\Delta \equiv 3$ (mod 8) *all classes g are properly primitive or all are improperly primitive according as $c_2(A)(-1)^r$ is* 1 *or* -1 *where r is the highest power of* 2 *in Ω and $c_2(A)$ is the Hasse invariant.*
3. *If* $\Delta \equiv 7$ (mod 8) *and Ω is even, all classes g are properly primitive or all are improperly primitive.*
4. *If* $\Delta \equiv 7$ (mod 8) *and Ω is odd, properly primitive and improperly primitive classes occur.*

Except in case 4, *all forms g associated with the genus of A are in the same genus.*

We showed in theorem 82 that if $\Delta \equiv 7$ (mod 8), the number of properly primitive classes is equal to the number of improperly primitive classes.

It is apparent that the determinant of any improperly primitive form is congruent to 3 (mod 4); this establishes 1. If $\Delta \equiv 3 \pmod 8$ the value of the Hasse invariant $c_2(g)$ is equal to the value of the Hilbert symbol $(-a, -\Delta)_2$. Since a may be taken odd or double an odd according as g is properly primitive or improperly primitive we see that $c_2(g) = 1$ or -1 in the respective cases and $c_2(E) = c_2(\Omega g) = (\Omega, -\Delta)_2 c_2(g)$. But $A \cong q \dotplus Eq^{-1} \cong q \dotplus Eq$ in $R(2)$ and hence $c_2(A) = (q, 1)_2 c_2(1 \dotplus E) = c_2(E) = (\Omega, -\Delta)_2 c_2(g)$. But $(\Omega, -\Delta)_2 = 1$ or -1 according as r is even or odd and condition 2. is established.

From theorem 37 we see that in 3. above, $A_1 \cong A_2$ in $R(2)$ implies $E_1' \cong E_2'$ in $R(2)$. Thus in cases 1., 2. and 3. above only one genus of forms g may occur for all forms of the genus of A.

It remains to consider $\Delta \equiv 7 \pmod 8$ with Ω odd. We have shown in the proof of 2. that $c_2(A) = c_2(g) = (-a, -\Delta)_2$. But $\Delta \equiv 7 \pmod 8$ implies that $(-a, -\Delta)_2 = 1$ whether or not a is odd or even. We need to show that a properly primitive form occurs if and only if an improperly primitive form occurs. Suppose $g = ax^2 + 2bxy + cy^2$ is properly primitive. We may take a to be double an odd integer and use the transformation

$$P = \frac{1}{2}\begin{bmatrix} 1 & \mp 3 \\ \pm 1 & 1 \end{bmatrix}$$

to take g into $g' = \frac{1}{4}(a'x^2 + 2b'xy + c'y^2)$ with $a' = a \pm 2b + c$, $b' = \mp 3a - 2b \pm c$, $c' = 9a \mp 6b + c$. Now $a \equiv 2 \pmod 4$ and $ac - b^2 \equiv 7 \pmod 8$ implies b odd and c divisible by 4; hence $a + 2b + c - (a - 2b + c) = 4b \equiv 4 \pmod 8$ with $a \pm 2b + c$ divisible by 4. Thus, for the ambiguous sign properly chosen, $a' \equiv 4 \pmod 8$ and g' is properly primitive. If on the other hand g is properly primitive we may take a to be odd and, by replacing x by $x + y$ if necessary make the coefficient of y^2 also odd; hence assume ac odd. Then the transformation P above with $\pm = +$ yields a form g'. But $ac - b^2 \equiv 7 \pmod 8$ and ac odd implies b even and $c \equiv 7a + ab^2$ or $a + c - b^2 \equiv a + 2b + c \equiv 0 \pmod 8$. Hence $a' \equiv c' \equiv 0 \pmod 8$ and g' is improperly primitive. The transformation P has denominator prime to p for all odd primes p and hence $c_p(g) = c_p(g')$ in both cases for all primes but $p = 2$. Furthermore $c_2(g) = c_2(g') = 1$ and hence $c_2(f_i) = 1$ for $f_1 = 1 \dotplus g$ and $f_2 = 1 \dotplus g'$. This shows that $f_1 \cong f_2$ in $R(2)$ from theorem 36. Thus $A_1 \text{ v } A_2$ and our proof is complete.

Recall from section 38 that λ_E is the number of distinct values mod q of the leading elements in the automorphs of E. First suppose E' is definite or a zero form. Then, from theorems 51a and 51b, the only automorphs are $\pm I$, where I is the identity matrix, except for $qd/\Omega^2 = 1$ or 3. But unless $q = \pm 1$ only the exception $\mathrm{d}/\Omega^2 = \pm 1$, $q = \pm 3$ need be considered and any automorph

$$\equiv \begin{bmatrix} w & 0 \\ u & v \end{bmatrix} \pmod 3$$

for integers w, u and v must have $w \equiv \pm 1 \pmod 3$. (The zero element occurs since we may consider $g \equiv -x^2 \pmod q$). Hence $\lambda_E = 2$ unless $q = \pm 1$ in which case $\lambda_E = 1$ and we have

$$(68) \qquad \begin{aligned} G(A, q) &= 2^{\mu(q)-1} h(E') &\quad \text{for } q \neq \pm 1 \\ &= h(E') &\quad \text{for } q = \pm 1 \end{aligned}$$

for $qd > 0$ or $-qd$ a square, where $h(E')$ is the number of classes in the genus of E' except that when $qd \equiv 7 \pmod 8$ we replace $h(E')$ by the sum of the number of classes in the two genera which occur.

Second, if E' is indefinite and non-zero, that is, if $-qd > 0$ and is not a perfect square, all automorphs of g are $\pm I$ multiplied by powers of

$$Q = \begin{bmatrix} t - bu & -cu \\ au & t + bu \end{bmatrix}$$

where t, u is the fundamental solution of the Pell equation $x^2 + \Delta y^2 = 1$ and $\Delta = ac - b^2$; that is, of all the solutions of the Pell equation except the trivial one, t is positive and has the least absolute value. Since we showed in section 38 that we may consider $g \equiv -x^2 \pmod{q}$, the first row of Q is congruent to $(t\ 0) \pmod{q}$ and the first row of Q^2 is congruent to $(t^2\ 0) \equiv (1\ 0) \pmod{q}$. Hence $\lambda_E = 2$ or 4 according as the fundamental solution (t, u) of $x^2 + \Delta y^2 = 1$ has or has not $t \equiv \pm 1 \pmod{q}$. In the respective cases

$$(69) \qquad \begin{aligned} G(A, q) &= 2^{\mu(q)-1} h(E') \quad \text{or} \quad 2^{\mu(q)-2} h(E'),\ q \neq \pm 1 \\ &= h(E') \quad \text{for} \quad q = \pm 1 \end{aligned}$$

except that when $dq \equiv 7 \pmod{8}$ and Ω is odd two genera of E' must be included.

Now let $t(w)$ denote the number of odd prime factors of w and see that $t(\Delta) = t(q) + t(d/\Omega^2)$. Combining the results of theorems 75, 82, 85, (68) and (69) we have

THEOREM 86. *For* $d = |A|$, Ω *the* g.c.d. *of the 2-rowed minor determinants of* A *and* $\Delta = qd/\Omega^2$, *with* q *prime to* $2d$ *and different from* ± 1,

$$G(A, q) = 2^{-t(d/\Omega 2)} h(\Delta)\rho \quad \text{or } 0$$

where $t(w)$ *is defined above,* $h(\Delta)$ *is the number of properly primitive classes of positive or indefinite binary forms* $ax^2 + 2bxy + cy^2$ *of determinant* $\Delta = ac - b^2$ *and*

1. *If* $\Delta \equiv 1$ *or* $2 \pmod 4$ *or* $\equiv 4 \pmod 8$, *then* $\rho = \frac{1}{2}$.
2. *If* $\Delta \equiv 7 \pmod 8$ *with* Ω *odd, then* $\rho = 2$.
3. *If* $\Delta \equiv 7 \pmod 8$ *with* Ω *even, then* $\rho = 1$.
4. *If* $\Delta \equiv 3 \pmod 8$, $\rho = 1/3$ *unless one of the following holds, in which case* $\rho = 1$:
   i) $c_2(A)(-1)^r = 1$, *where* r *is the highest power of* 2 *in* Ω.
   ii) $\Delta = 3$.
   iii) $c_2(A)(-1)^r = -1$, $\Delta < 0$, *and, for the fundamental solution* t, u *of* $t^2 + \Delta u^2 = 4$, tu *is odd.*
5. *If* $\Delta \equiv 0 \pmod 8$, *then* $\rho = \frac{1}{4}$.
6. *If* $\Delta < 0$ *and the fundamental solution of* $t^2 + \Delta u^2 = 1$ *has a value of* t *not congruent to* $\pm 1 \pmod q$, *divide the value of* ρ *given above by* 2.
7. *If* $-qd$ *is a square, halve the value of* ρ

Furthermore

$$G(A, \pm 1) = 2^{-t(d/\Omega^2)} h(\pm d/\Omega^2) 2\rho$$

where ρ *has the same values as above except that condition* 6 *is here vacuous and if* $\mp d$ *is a square, halve the value of* ρ.

In fact, if $\mp d$ is a square, we know from corollary 77 that $h(\pm d/\Omega^2) = 2\sqrt{\mp d/\Omega^2}$ and

$$G(A, \pm 1) = 2^{-t(d/\Omega^2)+1} \rho \sqrt{\mp d/\Omega^2}.$$

Notice also that if A is a positive matrix, Δ must be positive and that Δ may be positive even if A is indefinite.

Now if we let $N_0(A, q)$ be the number of essentially distinct representations of q by A (primitive or not) and $G_0(A, q)$ the sum of $N_0(A, q)$ over all classes in the genus of A, we see that

$$G_0(A, q) = \sum_{\lambda^2 / q} G(A, q/\lambda^2). \tag{70}$$

With reference to theorem 86 notice that dividing q by an odd square does not alter Δ (mod 8) nor does it alter $c_2(A)$, r nor $t(d/\Omega^2)$. Hence, except for cases 4 iii) and 6 which involve the Pell equation, ρ is the same for each G in (72). That the former exception cannot be eliminated is shown by the fact that the fundamental solution of $t^2 - 21u^2 = 4$ is $t = 5$, $u = 1$ while the fundamental solution of $t^2 - 189u^2 = 4$ is $t = 110$, $u = 8$. The second exception must be reckoned with in view of the example: $t^2 - 3u^2 = 1$ whose fundamental solution is $t = 2$, $u = 1$; $t^2 - 75u^2 = 1$ whose fundamental solution is $t = 26$, $u = 3$ and for $q = 3$ or 75 in the respective cases $2 \equiv \pm 1 \pmod{q}$ but $26 \not\equiv \pm 1 \pmod{q}$. However, neither exception occurs if Δ is positive and in that case we have

$$G_0(A, q) = 2^{-t(d/\Omega^2)}\rho \sum_{\lambda^2/q} h(\Delta/\lambda^2) \quad \text{or } 0$$

and hence if d is square-free

$$G_0(A, q) = 2^{-t(d/\Omega^2)} F(\Delta)\rho \quad \text{or } 0$$

where $F(\Delta)$ is the number of classes of forms of determinant Δ with one odd coefficient and middle coefficient even, this to be multiplied by 2 if $q = \pm 1$.

BIBLIOGRAPHY

1. Dickson, L. E. *Studies in the Theory of Numbers,* Chicago, 1930. Chap. IV.
2. Dolciani, Mary P., *On the Representations of Integers by Quadratic Forms,* Cornell Thesis, 1947.
3. Eisenstein, G. Tabelle der reducirten positive ternären quadratischen Formen. *Journal für Math.*, vol. 41 (1851), 141–190.
4. Hasse, H. *Über die Darstellbarkeit von Zahlen den quadratische Formen in Korper der rational Zahlen.* Jour. für Math., vol. 152 (1923), 129–148; also 205–224.

5. Jones, B. W. *A Table of Eisenstein-reduced positive ternary quadratic forms of determinant* $\leqq$ 200. National Research Council Bulletin number 97, (1935).

6. Kloosterman, H. D. *On the representation of numbers in the form* $ax^2 + by^2 + cz^2 + dt^2$, Acta Mathematica, vol. 49 (1926), 407–464.

7. Pall, Gordon. *The completion of a problem of Kloosterman*, Amer. Jour. Math., vol. 68 (1946), pp. 47–58.

8. Pall, Gordon and Jones, B. W. *Regular and semi-regular positive ternary quadratic forms*, Acta Math., vol. 70 (1940), 165–191.

9. Pall, Gordon and Ross, A. E. *The extension of a problem of Kloosterman*, Amer. Jour. Math., vol. 68 (1946), 59–65.

10. Siegel, C. L. *Über die analytische Theorie der quadratischen Formen*, Ann. Math., vol. 36 (1935), 527–606.

11. Siegel, C. L. ibid. II, vol. 37 (1936), 230–263.

12. Siegel, C. L. ibid. III, vol. 38 (1937), 212–291.

13. Siegel, C. L. *Einheiten quadratischen Formen*, Abh. Math. Sem. Hansischen Univ., vol. 13 (1939), 209–239.

14. Siegel, C. L. *Equivalence of quadratic forms*, Amer. Jour. Math., vol. 63 (1941), 658–680.

15. Tartakowsky, W. A. Compte Rendus de l'Academie des Sciences, vol. 186 (1928), 1337–1340, 1401–1403, 1684–1687. Errata in second paper are corrected in vol. 187, 155. Complete paper in Bull. Ak. Sc. U.R.S.S. (7) (1929), 111–22, 165–96.

PROBLEMS

Chapter I 1–8

1. By substitution find the form g into which

$$f = 8x_1^2 + 3x_2^2 + 6x_3^2 - 6x_1x_2 - 12x_1x_3 + 2x_2x_3$$

is taken by the transformation:

$$x_1 = -y_1 + y_2 + y_3\,, \quad x_2 = -y_1 + y_3\,, \quad x_3 = -y_1 + y_2\,.$$

Check your result by matric multiplication.

2. Using the methods of the proof of theorem 1, reduce the form f above to the form g of theorem 1 and hence find the canonical form of f under a real transformation.

3. Transform the following into a regular form:

$$2x_1x_3 + 4x_2x_4 + x_3^2 + x_4^2$$

and find its index using theorem 4.

4. Does the form in problem 3 represent the form in problem 1 in the field of real numbers?

5. Find a non-singular 3 by 3 matrix whose first two rows are (312) and (623).

6. Where, in the proof of theorem 8 is the condition $2 \neq 0$ used?

7. What is a necessary and sufficient condition that two forms with complex coefficients be congruent under a complex transformation?

8. What is a condition that one form will represent another form in the field of complex numbers?

Chapter II 9–30

9. Find 3-adic numbers which satisfy the following equations:

$$5x + 1 = 0, \quad x^2 = 7, \quad 45x = 1.$$

10. It is known that the decimal expansion of any rational number repeats and that any repeating decimal represents a rational number. State and prove the analogous theorem for any p-adic expansion; what can be said about the maximum length of the repeating part?

11. Prove the associative property of the multiplication of p-adic numbers.

12. Find the first three terms in a solution in 5-adic integers of $x^2 + 6y^2 = 5$.

13. For what values of a and b is the equation $ax^2 + by^2 = 6$ solvable in 2-adic integers?

14. For what values of a is $5x^2 = a$ solvable in 7-adic integers?

15. Prove that if b is an integer not divisible by p^2 and c is an integer congruent to $b \pmod{p^5}$ then $cx^2 = b$ for some p-adic integer x.

16. If p is an odd prime and n a non-square unit in $R(p)$, show that for every non-square unit n' in $F(p)$, the equation $x^2n' = n$ is solvable. What is the corresponding result for $p = 2$?

17. Find $c_p(f)$ for $f = \alpha_1 x_1^2 + \alpha_2 x_2^2$ where α_1, α_2 are p-adic integers.

18. Find $c_p(f)$ for $f = \alpha_1 x_1^2 + \alpha_2 x_2^2 + \alpha_3 x_3^2 + \alpha_4 x_4^2$ where α_i are p-adic integers. Compare section 11 for ternary forms.

19. Prove property 4 of the Hasse symbol and property 5 of $k_p(f)$.

20. Find $k_p(f)$ for $f = \alpha_1 x_1^2 + \alpha_2 x_2^2 + \alpha_3 x_3^2$.

21. Find an explicit formula for $c_p(f_1 + pf_2)$ in terms of the Hasse invariants and determinants of f_1 and f_2 and the number of variables in the two forms.

22. If $f = 3x_1^2 - 2x_2^2 - x_3^2$, it is a zero form in the field of rational numbers. Find a rational solution of $f = 5$. For what primes p is your solution in $R(p)$?

23. In what p-adic fields are the following forms zero forms?

$$f_1 = 5x_1^2 - x_2^2 - 3x_3^2, \quad f_2 = x_1^2 + x_2^2 + 7x_3^2 + 5x_4^2 .$$

24. In what p-adic fields are the following solvable: $3x_1^2 - x_2^2 = 5, f_1 = 3, f_2 = 4$ where the latter two forms are defined in the previous problem.

25. If f is a ternary form for which $c_p(f) = 1$, show that f represents all numbers N in $F(p)$.

26. Show that for $n < 4$, theorem 14 and corollary 14 hold for $p = \infty$.

27. In what p-adic fields are the following pairs of forms congruent:

a) $f_1 = 3x_1^2 + 7x_2^2$ and $f_2 = x_1^2 + 84x_2^2$.
b) $g_1 = x_1^2 - 3x_2^2 + 15x_3^2$ and $g_2 = 3x_1^2 - 5x_2^2 + 3x_3^2$.
c) $h_1 = x_1^2 - 5x_2^2 + 3x_3^2 - 7x_4^2$ and $h_2 = x_1^2 - x_2^2 + x_3^2 - x_4^2$.

28. For what primes p does g_1 represent f_1, h_1 represent f_1, using the notations of the previous problem?

29. Why does theorem 17 fail to hold for $p = \infty$? Give an example.

30. For what primes p are h_1 and h_2 in problem 27, 2-zero forms? For what p are they universal?

Chapter III 31–48

31. Find the Hermite reduced positive ternary forms with integral coefficients of determinants 2 and 3. Find the Hermite reduced binary forms with integral coefficients and determinant -2.

32. How many reduced binary forms are there of determinant 1 whose coefficients are rational numbers whose denominators divide 12?

33. If T is the matrix

$$\begin{bmatrix} 3 & 1 & 2 \\ 4 & 0 & 6 \\ 1 & 2 & 3 \end{bmatrix}$$

reduce it by matrices of type E_1 and E_2 in lemma 4 to the diagonal form of the lemma.

34. Show how the result of the previous problem may be achieved by use of lemma 5 without actually carrying through the reduction.

35. Find a column matrix T_0 such that the matrix $(T \quad T_0)$ is unimodular where T consists of the first two columns of problem 33.

36. If f is an integral zero form of square-free determinant prove that it is equivalent to a form whose matrix is

$$\begin{bmatrix} N & I_m & N_1 \\ I_m & D & N_1 \\ N_1 & N_1 & C \end{bmatrix}$$

where I_m is the m-rowed identity matrix, N_1 is the zero matrix with $n - 2m$ columns and m rows, C is the matrix of a non-zero form and D is a diagonal matrix each of whose elements is either 1 or 0.

37. Given $f = 3x^2 + 2xy + 4y^2$ find a form equivalent to f and congruent to $a_1x^2 + a_2y^2 \pmod{25}$ for properly chosen integers a_1 and a_2.

38. Prove that if f is a binary form with integral coefficients and determinant d and if for some prime q, $f = q$ is solvable in $F(p)$ for $p = \infty$ and all p dividing $2d$, then $f = q$ is solvable in $F(p)$ for all primes p.

39. Show that $x^2 + 3y^2 = 55$ is solvable in $F(2)$, $F(3)$, $F(\infty)$ but not in $F(5)$ nor $F(11)$. Find a number $M \equiv 55$

(mod 72) such that the equation is solvable in $F(p)$ for all primes p.

40. Prove that every binary zero form in the field of rationals is the product of two linear forms.

41. Which of the following are zero forms in the field of rationals:

$$3x^2 + 5y^2 - 7z^2, 2x^2 - 3y^2 + 11z^2, 2x^2 - 3y^2 + 5z^2 + 7t^2,$$
$$x^2 + 3y^2 - 7z^2 + 51t^2?$$

Where the answer is "yes", find values of the variables not all zero making the form zero.

42. What numbers are represented rationally by the forms in the previous problem?

43. Prove: if $p \equiv 3 \pmod 4$ and is a prime, while M is a rational number in lowest terms, then

$$2 + pM^2 = x^2 + y^2 - pz^2$$

is solvable for rational numbers x, y and z if and only if the denominator of M is prime to $2p$.

44. Prove: if p and q are odd primes with $p \equiv 1 \pmod 4$ and $(q \mid p) = -1$, while M is a rational number in lowest terms, then

$$2 + pqM^2 = x^2 + qy^2 - pz^2$$

is solvable for rational numbers x, y, z if and only if the denominator of M is prime to pq.

45. Fill in the details of the proof of Corollary 27c.

46. Find a form, different from that given at the close of section 21, which is universal in the field of rational numbers but is not a zero form in that field.

47. Show that the following two forms are rationally congruent:

$y_1^2 + y_2^2 + 16y_3^2$ and $2x_1^2 + 2x_2^2 + 5x_3^2 - 2x_2x_3 - 2x_1x_3$.

48. Construct an integral ternary form f of determinant 6 and index 2 such that $c_3(f) = -1 = c_2(f)$.

Chapter IV 49–57

49. Find a diagonal form with integral coefficients equivalent in $R(5)$ to $3x^2 + 2xy + 5y^2$. Can this be done in $R(2)$?

50. Find a form of the type of theorem 33*a* equivalent to

$x_1^2 + 2x_2^2 + 6x_2x_3 + 6x_3^2$; to $2x_1^2 + 6x_1x_2 + 6x_2^2 + 6x_3x_4$.

51. Given an improperly primitive form f of unit determinant. To what forms of the types in theorem 33*a* may it be equivalent? How may one determine for a given form to which type of form it will reduce?

52. Establish the statement made in the remark after theorem 34.

53. If f is a ternary form in $R(p)$ of non-zero determinant d and p is an odd prime factor of d, while p^2 does not divide d, what numbers N are represented by f in $R(p)$?

54. What odd numbers are represented in $R(2)$ by the forms in problem 50 above?

55. In what rings $R(p)$ are the following two forms equivalent?

$x_1^2 + 2x_2^2 + 6x_2x_3 + 6x_3^2, \quad 2x_1^2 + 6x_1x_2 + 6x_2^2 + 7x_3^2$.

56. Fill in the details of the proof of theorem 38 for $r = 3$ and $r = 1$.

57. Are the following two forms equivalent in $R(2)$?

$$\begin{aligned} f &= 6x_1x_2 + 2x_3^2 + 6x_4^2 + 8x_5^2 + 8x_5x_6 + 16x_6^2 , \\ g &= 2x_1x_2 + 2x_1x_3 + 2x_2^2 + 4x_2x_3 + 4x_3^2 + 6x_4^2 + 12x_4x_5 + \\ &\quad 14x_5^2 + 8x_5x_6 + 8x_6^2 . \end{aligned}$$

Chapter V 58–64

58. Given the matrix $T = \begin{bmatrix} 1 & 0 & 3 \\ 2 & 4 & 0 \\ 1 & 3 & 5 \end{bmatrix}$,

whose determinant is congruent to -1 (mod 3). Find a matrix $S \equiv T$ (mod 3) whose determinant is -1.

59. The matrices $T_1^T = (1, 2, 0)$ and $T_2^T = (1, 1, 1)$ are solutions of $T^TAT = 5$ where A is the matrix of the form $x^2 + y^2 + 3x^2$. Use lemma 10 to find an automorph M of A with rational elements such that $T_1 = MT_2$.

60. Find the numbers represented by $x^2 + 2y^2 + 3z^2$.

61. Meyer's theorem (theorem 83) shows that the form $x^2 + 2y^2 - 6z^2$ is in a genus of one class. What numbers does it represent?

62. Let f have F as its matrix and adj f be the form whose matrix is $F^{-1}|f|$. Prove that f primitively represents a binary form of determinant q, where f is a ternary form, if and only if adj f represents q primitively.

63. What binary forms are represented by $x^2 + 2y^2 + 3z^2$? What by $x^2 + 2y^2 - 6z^2$?

64. Find a properly primitive positive ternary form f with integral coefficients and of determinant 15 such that $c_3(f) = c_5(f) = 1$.

Chapter VI 65, 66

65. Let $n = 3$, $m = 1$, $A = \text{I}$, $B = 5$. Find a set of forms $\mathfrak{G}$ as described in section 38.

66. If $m = 2$ it is shown in theorem 51a that if $dq^{n-3} > 1$, the only automorphs of E_i are $\pm\text{I}$. What simplification does this introduce into theorem 47b?

Chapter VII 67–92

67. Prove that if p is an odd prime and a transformation T of determinant congruent to 1 (mod p) takes f into a form congruent to f (mod p) then T is expressible in the form P_0 of the theorem 50 where t, u is a solution of $x^2 + dy^2 \equiv 1$ (mod p). Would the same result hold if p were a power of a prime or, more generally, a composite number?

68. What modification of theorem 51a and 51c would be necessary if the restriction that the form be primitive were omitted?

69. Find the fundamental solutions of the Pell equations:

$$x^2 - 3y^2 = 1, \quad x^2 - 2y^2 = 1.$$

70. Find all the automorphs (proper and improper) for the following forms:
a) $x^2 + dy^2$ where d is an integer (positive, negative),
b) $3x^2 + 3xy + 5y^2$,
c) $3x^2 + 3xy - 5y^2$.

71. Find $M(d, q)$ and $M_0(d, q)$ for all values of q and binary forms having the following values of d: 1, -1, 2, -2, 3, -3.

72. Why, in the beginning of the proof of theorem 53 may f be written in the form given for f_0 even when $p = 2$?

73. Find the number of solutions of $3x^2 + 16xy + 5y^2 = 2^k q$ for q odd and various positive integer values of k. Find the number of solutions of $3x^2 + 13xy + 10y^2 = 130$.

74. Prove that the roots of the equation $x^2 + rx + s = 0$, where r and s are integers, are in $J(\Delta)$ as defined in section 42 with $\Delta = r^2 - 4s$. Conversely every quadratic integer α is a root of an equation $x^2 + rx + s = 0$ where r and s are integers, s being the norm of α.

75. Prove that $J(\Delta)$ is closed under addition, subtraction and multiplication. Is it closed under division?

76. A quadratic integer whose reciprocal is a quadratic integer is called a unit. Prove that a quadratic integer is a unit if and only if its norm is ± 1. Show that the number of units for every Δ is finite if $\Delta < 0$ and infinite if $\Delta > 0$, where Δ is a non-square integer.

77. Show that if $\Delta = 5$, every number in $J(\Delta)$ is in the ideal $(3, 1 + \sqrt{5})$.

78. Find a basis of the ideal $(6, 1 + 2\sqrt{7})$. What is its norm?

79. Find a basis for the product of the ideal in the previous problem and $(9, -1 + 2\sqrt{7})$. Show that theorem 61 holds for these two ideals.

80. Find a quadratic form associated with the ideal $[2, 1 + \sqrt{5}]$.

81. Find an ideal associated with the quadratic form $2x^2 + 3xy + 7y^2$.

82. Given the forms $f = 2x^2 + 15y^2$ and $g = 6x^2 + 5y^2$, find a form obtained from them by composition.

83. Prove theorem 67 for any n-ary primitive form.

84. Show by an example that theorem 68 does not apply to ternary forms.

85. Why is it necessary in theorem 71 to specify "proper" classes?

86. If Γ is the class of $2x^2 + 15y^2$ find Γ^2.

87. Let Γ be the principal class represented by $4x^2 + 9y^2$. Find a class Γ_1 such that $\Gamma_1^2 = \Gamma$.

88. If p/q is the fraction in lowest terms equal to $\{a_1, a_2, \cdots, a_k\}$ prove that $p = p_{k-1}a_k + p_{k-2}$ and $q = q_{k-1}a_k + q_{k-2}$ where p_{k-1}/q_{k-1} and p_{k-2}/q_{k-2} are the $k-1$st and $k-2$nd convergents, respectively, of p/q. Notice that this result does not follow immediately from lemma 18.

89. Using theorem 76 find all positive definite reduced binary forms of determinants $23/4$ and 11. Arrange them according to genera and find the number of classes in each genus, thus checking theorems 75 and 71.

90. Justify the remarks in the paragraph after corollary 76.

91. Complete the detail for $t = 0$ in the second paragraph of the proof of theorem 77.

92. Find all the classes of indefinite binary forms of

determinant -23, of determinant -15. Is there more than one class in any genus for either of these determinants?

Chapter VIII 93–95

93. Find the progressions associated with the genus of the form $f = x^2 + y^2 + 3z^2$. Show by theorem 23 that there is only one class in the genus of f.

94. Find the progressions associated with the genus of $f = 2x^2 + 6y^2 + 9z^2$.

95. Each form below is in a genus of one class. Verify the results of theorem 86 for the following forms and integers q.

a) $f = x^2 + y^2 + z^2$, $q = 5, 3, 11$.
b) $f = x^2 + y^2 + 2z^2$, $q = 5$.
c) $f = x^2 + y^2 + 4z^2$, $q = 5$.
d) $f = x^2 + 2y^2 + 2z^2$, $q = 7$.
e) $f = x^2 + y^2 + 3z^2$, $q = 5, 9, 17$.
f) $f = x^2 + y^2 + 8z^2$, $q = 5$.
g) $f = x^2 + y^2 + 5z^2$, $q = 7$.
h) $f = x^2 + 4y^2 + 12z^2$, $q = 5$.

To aid in the computations involved we append the following table of class numbers

n	3	5	10	11	15	20	35	40	51
$h(n)$	1	2	2	3	2	4	6	4	4.

THEOREM INDEX

INDEX